Building Regulations 2000

Approved Document B: Fire Safety (Volume 2) – Buildings other than dwellinghouses

Incorporating Insurers' Requirements for Property Protection

Published by RIBA Publishing, 15 Bonhill Street,
London EC2P 2EA

ISBN 978 1 85946 315 4

Stock Code 67167

British Library Cataloguing in Publications Data
A catalogue record for this book is available from the British Library.

RIBA Publishing and NBS are part of RIBA Enterprises Ltd. www.ribaenterprises.com

Building Regulations 2000

Approved Document B: Fire Safety (Volume 2) – Buildings other than dwellinghouses

Incorporating Insurers' Requirements for Property Protection

RIBA Publishing
RISCAuthority formerly InFiReS
FPA Fire Protection Association
Communities and Local Government

Foreword

Being an architect is to juggle the demands of regulatory control and best practice without losing sight of the client's brief, your design vision and the wider needs of society. This is a challenge which is fascinating and frustrating by equal measure, and ultimately tremendously rewarding. The frustrations arise from the sheer volume of regulation and guidance – not just Part B – that impact on design freedom and the onerous responsibility to 'stay on top of it' as rules change and sometimes conflict with each other. Of course, not to stay on top of it is both unprofessional and potentially dangerous.

Fortunately, this new guide helps to minimize some of these frustrations by aligning the sometimes overlooked guidance for property protection with the much more familiar Approved Document B all in one place. The concept of succinctly and transparently annotating the existing document is a practical and easy-to-use shortcut to meeting insurer's requirements at the same time as satisfying those of the Building Regulations. It has other benefits, too, not least in providing an enhanced service to clients, many of whom will understand the advantages of building-in measures to protect property from fire.

The FPA with RISCAuthority (formerly InFiReS) are to be commended both for this straightforward annotated version of Part B and for the enlightened inclusion of the 'Insurer requirements for the implementation of fire safety engineering solutions' in the Appendix, an equally useful document. Fire safety is an essential part of a designer's considerations; any publication helping to facilitate this is to be applauded. Ensuring that a client's asset is safeguarded must also be at the top of any professional's agenda. The information assembled in this book addresses both these issues. Architects and, indeed, all construction professionals and other stakeholders in the design of new commercial buildings, will I hope find this document a valuable addition to their libraries and a useful tool for designing more safely.

Peter Caplehorn
RIBA Practice Committee & Technical Director, Scott Brownrigg

Preface

According to the UK government's own figures, fire costs the UK economy in excess of £7 billion every year, yet despite this, those with an interest in property protection appear to have consistently failed to convey to the design community their message promoting the importance of inbuilt fire protection in buildings.

In 1999, on behalf of the insurers, The Fire Protection Association published the **Design Guide for the Fire Protection of Buildings**, which attempted to outline comprehensive guidance to allow architects and designers to take account of business risk management issues in relation to the fire protection of buildings. The publication was acclaimed for being clear and concise, it was referenced in Approved Document B and was made freely available to designers throughout the UK, but the principles it espouses have not been as widely embraced as the authors or insurers would have desired. In reality, fire, whilst remaining a major risk for commercial insurers, is only a minor issue for designers and building owners. Statistics for business failures following a major fire however do remain worryingly high, and so it remains in the best interests of all stakeholders that the principles of property protection are fully understood by all.

The **Essential Principles**, published through InFiReS, aims to provide an insurance industry equivalent to the basic fire safety objectives – the functional requirements – which are contained in Part B of Schedule 1 to the Building Regulations. The document presented here takes the logical next step. Using Approved Document B as the base, it simply amends or adds the requirements for property protection and business resilience. This will provide the design community with a single point of reference in a format with which they are familiar, so as to alert them where and when insurers may look for additional measures. It is well known that building in these measures as early in the design process as possible can considerably reduce the cost of implementation. We hope that this approach will ensure the design community have an improved understanding of the principles of property protection, and by embracing these requirements, considerably improve community risk management and reduce the cost of fire to their clients and the economy of the UK.

Chris Hanks
Chairman of RISCAuthority (formerly InFiReS)

Preface

Having been involved with the construction community for many years, both inside and outside government, I have become acutely aware of the perceived lack of common ground between those who wish for more freedom to design, and those with an interest in property protection – principally the insurers.

Building regulations quite correctly focus on life safety, ensuring that the occupants of a building are safely evacuated in the event of fire. However, it is widely recognised that the functional requirements contained in Part B of Schedule 1 of the Building Regulations differ from those required for property protection. The publication of the **Essential Principles** document by InFiReS was broadly welcomed by all those with an interest in design and construction, since for the first time, the insurance industry simply and concisely stated their aims and objectives in a format familiar to the design community. This was reflected by their reference in the latest iteration of Approved Document B.

Nevertheless, although architects and designers now had free access to insurers, functional requirements, much of the detail was risk dependent and contained within a number of different references sources. So the decision taken by RISCAuthority (formerly InFiReS) to begin with the Approved Document and amend or insert additional requirements should be recognised as a major change in approach by the insurers' and should further bridge the gap in understanding between themselves and the design community. On behalf of the Fire Protection Association, I would like to thank the Department for Communities and Local Government (CLG) and the Royal Institution of British Architects (RIBA) for their assistance with this project; their involvement demonstrates the potential of partnership working and has been a major contributory factor in making the publication of this document possible.

New methods of property protection will continue to evolve as more innovative methods of construction and building materials emerge, and as the Approved Document B is reviewed over time so this publication will need to continue to be updated. Insurers, however, appreciate that designers increasingly require greater freedoms and we are experiencing a more widespread use of alternative solutions. The inclusion of *The Insurer requirements for the implementation of fire safety engineering solutions* as an appendix within this book recognises the move away from prescription to a performance based approach.

As society becomes more conscious of the need to reduce energy and emissions it will require greater efficiencies in the construction process, embracing new methods and materials. Understanding how these perform in fire is important for insurers and designers alike, and designing in and producing structures with greater resilience would appear to be in everyone's best interest. I hope that this publication can assist the process.

Rt Hon Nick Raynsford MP
Chairman of FPA Council

Introduction to RISCAuthority (formerly InFiReS)

Introduction and use of this guide

This document uses the text from **Approved Document B: Fire Safety (Volume 2) – Buildings other than dwellinghouses** and includes additional text, tables and diagrams to cover relevant property protection and business interruption issues. **Approved Document B** focuses primarily on life safety issues. These additions give guidance in relation to business and property protection – areas which are increasingly viewed as requirements for resilient and healthy communities. Appendix J includes an additional guide entitled *Insurer requirements for the implementation of fire safety engineering solutions.*

RISCAuthority

The additional text relating to property protection has been developed through the Risk Insight, Strategy and Control Authority (RISCAuthority), an insurer funded research initiative, and published by the Fire Protection Association (FPA). RISCAuthority membership comprises a group of UK insurers that actively support a number of expert working groups developing and promulgating best practice for the protection of property and business from loss due to fire and other risks. The technical expertise for this document has been provided by the Technical Division of the FPA, external consultants and experts from the insurance industry, who together form the various RISCAuthority Steering Groups. Although produced with insurer input, it does not (and is not intended to) represent a pan-insurer perspective. Individual insurance companies will have their own requirements, which may be different from, or not reflected in, the content of this document.

Aim

The aim of this document is to provide loss prevention guidance for those who design, construct and equip industrial and commercial buildings. This guide continues the long tradition of providing authoritative guidance on loss prevention issues started by the Fire Offices' Committee of the British insurance industry over one hundred years ago, and builds upon more recent publications in the *Design guide* series from the Loss Prevention Council, Association of British Insurers and the Fire Protection Association.

How to use this guide

As stated above, this guide reproduces in its entirety the text from **Approved Document B: Fire Safety (Volume 2) – Buildings other than dwellinghouses**, which will be familiar to those working in the design, construction and associated equipment industries.

Additional requirements for property protection designed to help you add resilience to your business are provided as white text on a green background in normal font; additional explanatory text and commentary is similarly highlighted, but appears in an italicised font.

In the case of tables and diagrams, wherever possible additional information has been incorporated with the original information for ease of use. However, certain tables and diagrams are repeated to ensure the clarity of the information provided.

For ease of use, an index of the key additional guidance has been included immediately following this Introduction referencing both relevant paragraph and page numbers.

Approved Document B allows for the 'engineering' of alternative building designs so long as equivalency in life safety performance is demonstrated. The guide *Insurer requirements for the implementation of fire safety engineering solutions* given in Appendix J provides an alternative set of design targets pertinent to business and property protection for fulfilment through 'normal' fire engineering design methods.

Index to additional guidance

The value of engaging with insurers at the earliest opportunity is also highlighted.

Para	Subject	Change/Addition
0.14	Property protection	Four objectives for property protection and twelve essential principles to be achieved in the design and construction of commercial and industrial premises. Commentary on use of fire suppression and additional compartmentation in education establishments.
0.16	Sprinkler systems	Compliance requirements of sprinkler systems; recommendation to seek advice from insurers.
0.21	Alternative approaches	Reference to FPA sector guides.
0.32	Alternative approaches – fire safety engineering	Additional fifth factor to be taken into account: the anticipated property damage and business interruption.
B1	The requirement: means of warning and escape	Three essential principles for property protection.
1.30	Buildings other than flats – fire alarm systems	Note on the potential requirement for enhanced property protection for insurance purposes and especially – subject to the fire risk assessment – large/complex risks.
1.39	Buildings other than flats – property protection	Insurers' requirements for fire detection systems in sprinklered and unsprinklered buildings.
B2	The requirement: internal fire spread (linings)	Three essential principles for property protection.
B2.iv	Introduction – furniture and fittings	Contribution made by contents of buildings to the development of fire, and resulting recommendations.
Table 10a	Classification of linings (property protection)	Table 10a provides further details on surface properties of linings with respect to property protection.
6.4	Variations and special provisions – walls	Reference to Table 10a.
6.7	Variations and special provisions – rooflights	Reference to Table 10a. Specification of plastic rooflights in sprinklered buildings.
6.11 Note 3	Thermoplastic materials – windows and internal glazing	Specifications for cavity barriers around plastic window frames. Commentary added on the use of combustible insulants.
6.12	Thermoplastic materials – rooflights	Extension to areas which may not have rooflights constructed of thermoplastic material; namely, those protected by sprinklers.

Index to additional guidance

Para	Subject	Change/Addition
Table 11 Notes	Limitations applied to thermoplastic rooflights and lighting diffusers in suspended ceilings and Class 3 plastic rooflights	Note on the spacing of rooflights.
B3	The requirement: internal fire spread (structure)	Three essential principles for property protection.
B3	Performance	Reference to the *FPA Design guide: Compartmentation*.
7.4	Fire resistance standard – exclusions from the provisions for elements of structure	Potential problem with roof collapse in single-storey buildings, potentially affecting a compartment or separating wall.
7.8 Note 4	Raised storage areas	Specifications for permitted floor areas and fire resistance of raised floor areas.
8.1	Compartmentation: introduction	Third objective of compartmentation: containing fire in compartment of origin to assist in reducing property damage and minimising business interruption. Strategies for achieving this are outlined. Commentary outlining the importance of compartmentation for property protection and limiting business interruption.
8.6	Junctions	Commentary regarding fire resistance and weakness at junctions.
8.8	Buildings containing one or more atria	Strategies for containing fire where an atrium breaches any compartmentation.
8.18	Provision of compartmentation – non-residential buildings	Addition of requirement for all floors to be compartment floors and reductions to allowable areas of each floor in certain circumstances. Preferred, reduced upper storey height detailed as trigger requirement for compartment floor construction.
Diagram 28	Compartment floors: illustration of guidance in paragraph 8.18	Additions to Diagram 28.
Table 12 Notes	Maximum dimensions of building or compartment (non-residential buildings)	Reference to Table 12a (see below).
Table 12a	Guidance on maximum dimensions of building or compartment (non-residential buildings) for property protection	Table 12a provides revised maximum dimensions of building or compartment and accompanying notes. A new diagram aids understanding of insurers' requirements. New diagram also outlines additional factors that insurers will take into account when recommending limits on floor area. Diagram to Table 12a added.
Table 12b	Maximum floor areas for specific applications in warehouses and storage buildings, subject to insurers' individual risk assessment	As a supplement to Table 12a.

Index to additional guidance

Para	Subject	Change/Addition
8.29 Note	Junction of compartment wall with roof	Additional restrictions placed on use of certain thermoplastic rooflights.
8.40	Pipes for oil or gas and ventilation ducts in protected shafts	Additional requirement for the protection of pipes containing volatile fluids.
9.1	Concealed spaces: introduction	Commentary on services that pass through cavity barriers and the specification of the cavity barrier system.
Table 13	Maximum dimensions of cavities in non-domestic buildings (Purpose Groups 2–7)	Addition of dimensions for a third location of cavity: cavities deeper than 600mm or above areas of special risk.
9.12	Extensive cavities – maximum dimensions of concealed spaces	Addition of considerations in respect to cables in communication rooms.
9.13a	Construction and fixings for cavity barriers	Supplementary requirement for additional insulation.
10.6	Openings for pipes – Alternative A: proprietary seals (any pipe diameter)	Commentary outlining importance of ensuring that the test result is applicable to the particular service and the material from which the service is made.
Table 14 Notes	Maximum nominal internal diameter of pipes passing through a compartment wall/floor	Commentary confirming the conditions under which provisions in the table are acceptable. Requirement for a closing device to be fitted to plastic pipes and small service ducts which pass through a compartment wall, floor or cavity barrier.
10.7	Openings for pipes – Alternative B: pipes with a restricted diameter	Commentary confirming reasons why Alternative A is preferred to Alternative B.
10.10	Ventilation ducts, flues etc.	Commentary on specifying effective grease filters to reduce build-up of grease in kitchen extract ducts.
10.13 Note	Ventilation ducts, flues etc. – fire dampers	Additional limitation of fire dampers actuated only by fusible links.
10.19	Fire-stopping	Addition of methods by which the fire resistance of a fire-separating element when penetrated by a service shall be reinstated to the levels recommended in Table A1 in Appendix A. Reference to Intumescent Fire Seals Association (IFSA) code.
B4	The requirement: external fire spread	Two essential principles for property protection.
12.2	Introduction	Commentary on resistance to a fire caused by penetration of an external fire.
13.6	Boundaries – notional boundaries	Reference to Diagram 42a (see below).
Diagram 42a	Combustible goods stored between building and notional boundary	Diagram 42a (as supplement to Diagram 42) illustrates the concern insurers have when two buildings have combustible materials stored between the building and the notional boundary specified by regulators.
Diagram 43a	Property protection provisions for external surfaces of walls	Addition of Diagram 43a to supplement Diagram 43.

Index to additional guidance

Para	Subject	Change/Addition
13.21	Arson prevention	Addition of recommendations for consideration by those involved in the design and construction of buildings. Reference to the FPA's publication *The Prevention and Control of Arson*.
Table 16	Limitations on roof coverings	Reference to Table 16a (see below).
Table 16a	Property protection – limitations on roof coverings	Supplement to Table 16. Commentary on the risk from hot-work, which should be considered when specifying materials used in the construction of roofs.
Table 17 Notes	Class 3 (National class) or Class D-s3, d2 (European class) plastic rooflights: limitations on use and boundary distance	Addition of dimensions for property protection.
Table 18	TP(a) and TP(b) plastic rooflights: limitations on use and boundary distance	Addition of further areas which may be served by TP(a) rigid plastic rooflights.
Table 18 Notes	TP(a) and TP(b) plastic rooflights: limitations on use and boundary distance	Addition of dimensions for property protection. Further issues to consider with respect to sprinklers and plastic rooflights.
B5.iii	Insulating core panels	Reference to additional information in Appendix F.
Additional to Table 19, note Table 19a	Fire and rescue service vehicle access to single-storey commercial and industrial buildings not fitted with fire mains	Addition of Table 19a to supplement Table 19.
17.6	Provision of firefighting shafts	Addition of Table 4 from BS5588-5: 2004 giving recommendations for firefighting shafts.
Appendix A 2	Introduction	Commentary on the higher levels of performance which may be required by insurers.
Appendix A 3	Fire resistance	Addition of a further factor having a bearing on fire resistance; namely, protection of business assets.
Appendix A 4	Fire resistance	Addition of a further factor affecting the standard of fire resistance specified; namely, the insurers' need to consider the risk of fire occurring and its effects. Commentary on insurers' experience of inception risks in some purpose groups and the consequent need to assume high fire loads.
Appendix A 13 Note 2	Internal linings	Addition of note on classification in respect to property protection for lining materials.
Table A1	Specific provisions of test for fire resistance of elements of structure etc.	Additional provisions for floors, compartment walls, fire resisting construction, cavity barrier and casing around a drainage system. Reference to Table A9 with respect to external walls.
Table A2	Minimum periods of fire resistance	Additional minimum periods of fire resistance for various purpose groups of buildings, and accompanying notes.

Index to additional guidance

Para	Subject	Change/Addition
Table A3 Notes	Limitations on fire-protecting suspended ceilings	Commentary on specifying suspended ceiling type Z for property protection.
Table A4a	Limitations on the use of uninsulated 'fire-resisting' glazed elements enclosing areas of specific risk	Addition of Table A4a to supplement Table A4.
Table A9	Property protection from arson attack – specific provisions of test for fire resistance	Table with accompanying notes.
Table B1	Provisions for fire doors	Additional provisions for fire doors in a compartment wall and any door forming part of an enclosure.
Appendix E	Definitions	Addition of definition for 'high risk area'. Addition of definition for 'protected zone'. Addition of definition for 'special risk'.
Appendix F	Composite panels – insurers' view	Commentary on the significant losses that have occurred over the last decade associated with large fires in which composite panels have been a feature.
Appendix G 1	Fire safety information	Commentary added on making information available to insurers and fire authorities.
Appendix H	Publications	List of Fire Protection Association (FPA) publications. Reference to Intumescent Fire Seals Association (IFSA) code. List of Loss Prevention Standards.
Appendix I	Main changes to the 2006 edition	
Appendix J	*Insurer requirements for the implementation of fire safety engineering solutions*	Additional appendix

Contents

Tables

Use of guidance

THE APPROVED DOCUMENTS

This document is one of a series that has been approved and issued by the Secretary of State for the purpose of providing practical guidance with respect to the requirements of Schedule 1 to and Regulation 7 of the Building Regulations 2000 (SI 2000/2531) for England and Wales.

At the back of this document is a list of all the documents that have been approved and issued by the Secretary of State for this purpose.

The Approved Documents are intended to provide guidance for some of the more common building situations. However, there may well be alternative ways of achieving compliance with the requirements.

Thus there is no obligation to adopt any particular solution contained in an Approved Document if you prefer to meet the relevant requirement in some other way.

Other requirements

The guidance contained in an Approved Document relates only to the particular requirements of the Regulations which that document addresses. The building work will also have to comply with the Requirements of any other relevant paragraphs in Schedule 1 to the Regulations.

There are Approved Documents which give guidance on each of the other requirements in Schedule 1 and on Regulation 7.

LIMITATION ON REQUIREMENTS

In accordance with Regulation 8, the requirements in Parts A to D, F to K, N and P (except for paragraphs H2 and J6) of Schedule 1 to the Building Regulations do not require anything to be done except for the purpose of securing reasonable standards of health and safety for persons in or about buildings (and any others who may be affected by buildings or matters connected with buildings). This is one of the categories of purpose for which Building Regulations may be made.

Paragraphs H2 and J6 are excluded from Regulation 8 because they deal directly with prevention of the contamination of water. Parts E and M (which deal, respectively, with resistance to the passage of sound and access to and use of buildings) are excluded from Regulation 8 because they address the welfare and convenience of building users. Part L is excluded from Regulation 8 because it addresses the conservation of fuel and power. All these matters are amongst the purposes other than health and safety that may be addressed by Building Regulations.

MATERIALS AND WORKMANSHIP

Any building work which is subject to the requirements imposed by Schedule 1 of the Building Regulations should, in accordance with Regulation 7, be carried out with proper materials and in a workmanlike manner.

You may show that you have complied with Regulation 7 in a number of ways. These include the appropriate use of a product bearing CE marking in accordance with the Construction Products Directive (89/106/EEC)[1], the Low Voltage Directive (73/23/EEC and amendment 93/68/EEC)[2] and the EMC Directive (89/336/EEC)[3], as amended by the CE Marking Directive (93/68/EEC)[4], or a product complying with an appropriate technical specification (as defined in those Directives), a British Standard, or an alternative national technical specification of a Member State of the European Union or Turkey[5], or of another State signatory to the Agreement on the European Economic Area (EEA) that provides an equivalent level of safety and protection, or a product covered by a national or European certificate issued by a European Technical Approval Issuing body and the conditions of use are in accordance with the terms of the certificate.

You will find further guidance in the Approved Document supporting Regulation 7 on materials and workmanship.

---

1 As implemented by the Construction Products Regulations 1991 (SI 1991 No 1620)

2 As implemented by the Electrical Equipment (Safety) Regulations 1994 (SI 1994 No 3260)

3 As implemented by the Electromagnetic Compatibility Regulations 1992 (SI 1992 No 2372)

4 As implemented by the Construction Products (Amendment) Regulations 1994 (SI 1994 No 3051) and the Electromagnetic Compatibility (Amendment) Regulations 1994 (SI 1994 No 3080)

5 Decision No. 1/95 of the EC-Turkey Association Council of 22 December 1995

Independent certification schemes

There are many UK product certification schemes. Such schemes certify compliance with the requirements of a recognised document which is appropriate to the purpose for which the product is to be used. Products which are not so certified may still conform to a relevant standard.

Many certification bodies which approve such schemes are accredited by the United Kingdom Accreditation Service (UKAS).

Since the fire performance of a product, component or structure is dependent upon satisfactory site installation and maintenance, independent schemes of certification and accreditation of installers and maintenance firms of such will provide confidence in the appropriate standard of workmanship being provided.

Building Control Bodies may accept the certification of products, components, materials or structures under such schemes as evidence of compliance with the relevant standard. Similarly, Building Control Bodies may accept the certification of the installation or maintenance of products, components, materials or structures under such schemes as evidence of compliance with the relevant standard. Nonetheless, a Building Control Body will wish to establish, in advance of the work, that any such scheme is adequate for the purposes of the Building Regulations.

Technical specifications

Building Regulations are made for specific purposes including: health and safety, energy conservation and the welfare and convenience of people. Standards and technical approvals are relevant guidance to the extent that they relate to these considerations. However, they may also address other aspects of performance such as serviceability, or aspects which, although they relate to health and safety, are not covered by the Regulations.

When an Approved Document makes reference to a named standard, the relevant version of the standard is the one listed at the end of the publication. However, if this version of the standard has been revised or updated by the issuing standards body, the new version may be used as a source of guidance provided it continues to address the relevant requirements of the Regulations.

The appropriate use of a product which complies with a European Technical Approval as defined in the Construction Products Directive should meet the relevant requirements.

The Department intends to issue periodic amendments to its Approved Documents to reflect emerging harmonised European Standards. Where a national standard is to be replaced by a European harmonised standard, there will be a co-existence period during which either standard may be referred to. At the end of the co-existence period the national standard will be withdrawn.

INTERACTION WITH OTHER LEGISLATION

The Regulatory Reform (Fire Safety) Order 2005

The Fire Safety Order reforms the law relating to fire safety in non-domestic premises. Specifically it replaces the Fire Precautions (Workplace) Regulations 1997 and the Fire Precautions Act 1971. It imposes a general duty to take such fire precautions as may be reasonably required to ensure that premises are safe for the occupants and those in the immediate vicinity.

By virtue of the Order, the responsible person is required to carry out a fire risk assessment of their premises. This must be a suitable and sufficient assessment of the risks to which relevant persons are exposed for the purpose of identifying the general fire precautions they need to take to comply with the requirements under the Order.

Although these requirements are applicable to premises whilst in operation, it would be useful for the designers of a building to carry out a preliminary fire risk assessment as part of the design process. If a preliminary risk assessment is produced, it can be used as part of the Building Regulations submission and can assist the fire safety enforcing authority in providing advice at an early stage as to what, if any, additional provisions may be necessary when the building is first occupied.

Article 6 of the Order does exclude some premises such as certain mines, vehicles and land forming part of an agricultural or forestry undertaking. The Order applies to all non-domestic premises, which includes the common parts of block of flats and HMOs.

Guidance on the consultation procedures that should be adopted to ensure that the requirements of all enforcing authorities are addressed at Building Regulation Approval stage is contained in *Building Regulation and Fire Safety – Procedural Guidance*, published jointly by CLG and the Welsh Assembly Government.

There may be other Statutes enforced by the local authority or the fire and rescue authority that may be applied to premises of specific uses once they are occupied.

Houses in multiple occupation

This guidance may also be applicable to the design and construction of dwellings which are considered to be 'houses in multiple occupation' (HMOs), as defined in the Housing Act 2004, providing there are no more than six residents in any self-contained dwelling. The licensing of HMOs is typically overseen by the Local Authority who may require additional precautions over and above this guidance. Technical guidance on the assessment of hazards from fire and preventive measures for HMOs is contained in the Housing Health and Safety Rating System Operating Guidance issued in February 2006 (ISBN: 978 1 85112 846 4).

The Workplace (Health, Safety and Welfare) Regulations 1992

The Workplace (Health, Safety and Welfare) Regulations 1992 contain some requirements which affect building design. The main requirements are now covered by the Building Regulations, but for further information see: *Workplace health, safety and welfare, The Workplace (Health, Safety and Welfare) Regulations 1992, Approved Code of Practice and Guidance;* The Health and Safety Commission, L24; published by HMSO 1992 (ISBN: 0 11886 333 9).

The Workplace (Health, Safety and Welfare) Regulations 1992 apply to the common parts of flats and similar buildings if people such as cleaners, wardens and caretakers are employed to work in these common parts. Where the requirements of the Building Regulations that are covered by this Part do not apply to dwellings, the provisions may still be required in the situations described above in order to satisfy the Workplace Regulations.

The Construction (Design and Management) Regulations 2006

The purpose of this Approved Document is to provide guidance on the fire safety requirements for the completed building. It does not address the risk of fire during the construction work which is covered by the Construction (Design and Management) Regulations 2006 and the Regulatory Reform (Fire Safety) Order. HSE has issued the following guidance on fire safety in construction: Construction Information Sheet No 51 Construction fire safety; and HSG 168 Fire safety in construction work (ISBN: 0 71761 332 1).

When the construction work is being carried out on a building which, apart from the construction site part of the building, is occupied, the Fire and Rescue Authority is responsible for the enforcement of the 2006 Regulations in respect of fire. Where the building is unoccupied, the Health and Safety Executive is responsible for enforcement on the construction site.

The Construction Products Directive

The Construction Products Directive (CPD) is one of the 'New Approach' Directives, which seek to remove technical barriers to trade within the European Economic Area (EEA) as part of the move to complete the Single Market. The EEA comprises the European Community and those states in the European Free Trade Association (other than Switzerland).

The intention of the CPD is to replace existing national standards and technical approvals with a single set of European-wide technical specifications for construction products (i.e. harmonised European standards or European Technical Approvals). Any manufacturer whose products have CE marking showing that they are specified according to European technical specifications cannot have these products refused entry to EEA markets on technical grounds. In the UK, the CPD was implemented by the Construction Products Regulations, which came into force on 27 December 1991 and were amended on 1 January 1995 by the Construction Products (Amendment) Regulations 1994.

This document refers to and utilises within its guidance, a large number of British Standards, in relation to Codes of Practice and fire test methods (typically the BS 476 series of documents). In order to facilitate harmonisation and the use of the new technical specifications and their supporting European test standards, guidance is also given on the classification of products in accordance with those standards.

Guidance is given for the appropriate use and/or specification of a product to which one or more of the following apply:

1. a product bearing CE marking in accordance with the Construction Products Directive (89/106/EEC) as amended by the CE marking Directive (93/68/EEC);
2. a product tested and classified in accordance with the European Standards (BS EN) referred to in the Commission Decision 2000/147/EC[1] and/or Commission Decision 2000/367/EC[2];
3. a product complying with an appropriate technical specification (as defined in the Directives 89/106/EC as amended by 93/68/EEC).

The implementation of the CPD will necessitate a time period during which national (British) Standards and European technical specifications will co-exist. This is the so-called period of co-existence. The objective of this period of co-existence is to provide for a gradual adaptation to the requirements of the CPD. It will enable producers, importers and distributors of construction products to sell stocks of products manufactured in line with the national rules previously in force and have new tests carried out. The duration of the period of co-existence in relation to the European fire tests has not yet been clearly defined.

As new information becomes available and further harmonised European Standards relevant to this document are published, further guidance will be made available.

---

1 Implementing Council Directive 89/106/EEC as regards the classification of the reaction to fire (2000/147/EC) performance of construction products.

2 Implementing Council Directive 89/106/EEC as regards the classification of the resistance to fire (2000/367/EC) performance of construction products, construction works and parts thereof.

Designation of standards

The designation of 'xxxx' is used for the year referred to for standards that are not yet published. The latest version of any standard may be used provided that it continues to address the relevant requirements of the Regulations.

Commission guidance papers and decisions

The following guidance papers and Commission Decisions are directly relevant to fire matters under the Construction Products Directive:

Guidance paper G

The European classification system for the reaction to fire performance of construction products.

Guidance paper J

Transitional arrangements under the Construction Products Directive.

Commission Decision of 8 February 2000 (2000/147/EC) implementing Council Directive 89/106/EEC as regards the classification of the reaction to fire performance of construction products.

Commission Decision of 3 May 2000 (2000/367/EC) implementing Council Directive 89/106/EEC as regards the classification of the resistance to fire performance of construction products, construction works and parts thereof.

Commission Decision of 26 September 2000 (2000/605/EC) amending Decision 96/603/EC establishing the list of products belonging to Classes A 'No contribution to fire' provided for in Decision 94/611/EC implementing Article 20 of Council Directive 89/106/EEC on construction products.

Corrigenda – Corrigendum to Commission Decision 2000/147/EC of 8 February 2000 implementing Council Directive 89/106/EEC as regards the classification of the reaction to fire performance of construction products.

The publication and revision of Commission guidance papers and decisions are ongoing and the latest information in this respect can be found by accessing the European Commission's website via the link on the CLG website at: www.communities.gov.uk/buildingregs.

Environmental Protection

Requirements under Part B of the Building Regulations and the guidance in this Approved Document are made for the purpose of ensuring the health and safety of people in and around buildings.

The Environment Agency publishes guidance on the design and construction of buildings for the purpose of protecting the environment. This includes Pollution Prevention Guidelines (PPG18) on *Managing Fire Water and Major Spillages,* which seeks to minimise the effects of water run-off from fire-fighting. It is aimed at medium to large (and small, high-risk) commercial and industrial sites and sets out requirements for the construction of containment areas for contaminated water and such other measures.

It should be noted that compliance with the Building Regulations does not depend upon compliance with other such guidance.

General introduction

FIRE SAFETY VOLUME 2 – BUILDINGS OTHER THAN DWELLINGHOUSES

Scope

0.1 Approved Document B (Fire safety) has been published in two volumes. Volume 1 deals solely with dwellinghouses (see Appendix E and Building Regulation 2(1)), while Volume 2 deals with all other types of building covered by the Building Regulations.

Where very large (over 18m in height) or unusual dwellinghouses are proposed, some of the guidance in Volume 2 may be needed to supplement that given by Volume 1.

ARRANGEMENT OF SECTIONS

0.2 The functional requirements B1 to B5 of Schedule 1 of the Building Regulations are dealt with separately in one or more Sections. The requirement is reproduced at the start of the relevant Sections, followed by an introduction to the subject.

0.3 The provisions set out in this document deal with different aspects of fire safety, with the following aims.

> **B1:** To ensure satisfactory provision of means of giving an alarm of fire and a satisfactory standard of means of escape for persons in the event of fire in a building.
>
> **B2:** To ensure fire spread over the internal linings of buildings is inhibited.
>
> **B3:** To ensure the stability of buildings in the event of fire; to ensure that there is a sufficient degree of fire separation within buildings and between adjoining buildings; to provide automatic fire suppression where necessary; and to inhibit the unseen spread of fire and smoke in concealed spaces in buildings.
>
> **B4:** To ensure external walls and roofs have adequate resistance to the spread of fire over the external envelope and that spread of fire from one building to another is restricted.
>
> **B5:** To ensure satisfactory access for fire appliances to buildings and the provision of facilities in buildings to assist firefighters in the saving of life of people in and around buildings.

0.4 Whilst guidance appropriate to each of these aspects is set out separately in this document, many of the provisions are closely interlinked. For example, there is a close link between the provisions for means of escape (B1) and those for the control of fire growth (B2), fire containment (B3) and facilities for the fire and rescue service (B5). Similarly there are links between B3 and the provisions for controlling external fire spread (B4) and between B3 and B5. Interaction between these different requirements should be recognised where variations in the standard of provision are being considered. A higher standard under one of the requirements may be of benefit in respect of one or more of the other requirements. The guidance in the document as a whole should be considered as a package aimed at achieving an acceptable standard of fire safety.

0.5 In the guidance on B1 the provisions for flats are separated from those for all other types of building because there are important differences in the approach that has been adopted.

Appendices: provisions common to more than one of Part B's requirements

0.6 Guidance on matters that refer to more than one of the Sections is in a series of Appendices, covering the following subjects:

- Appendix A: Performance of materials, products and structures
- Appendix B: Fire doors
- Appendix C: Methods of measurement
- Appendix D: Purpose groups
- Appendix E: Definitions
- Appendix F: Fire behaviour of insulating core panels used for internal structures
- Appendix G: Fire safety information
- Appendix H: Standards and other publications referred to.

Fire performance of materials, products and structures

0.7 Much of the guidance throughout this document is given in terms of performance in relation to standard fire test methods. Details are drawn together in Appendix A to which reference is made where appropriate. In the case of fire protection systems, reference is made to standards for systems design and installation. Standards referred to are listed in Appendix H.

Fire doors

0.8 Guidance in respect of fire doors is set out in Appendix B.

Methods of measurement

0.9 Some form of measurement is an integral part of much of the guidance in this document and methods are set out in Appendix C.

Purpose groups

0.10 Much of the guidance in this document is related to the use of the building. The use classifications are termed purpose groups and they are described in Appendix D.

Definitions

0.11 The definitions are given in Appendix E.

Fire safety Information

0.12 Regulation 16B requires that where building work is carried out which affects fire safety, and where the building affected will be covered by the Regulatory Reform (Fire Safety) Order 2005, the person carrying out the work must provide sufficient information for persons to operate and maintain the building in reasonable safety. This information will assist the eventual owner/occupier/employer to meet their statutory duties under the Regulatory Reform (Fire Safety) Order.

The exact amount of information and level of detail necessary will vary depending on the nature and complexity of the building's design.

For small buildings, basic information on the location and nature of fire protection measures may be all that is necessary.

For larger buildings, a more detailed record of the fire safety strategy and procedures for operating and maintaining any fire protection measures of the building will be necessary. Appendix G provides advice on the sort of information that should be provided.

MANAGEMENT OF PREMISES

0.13 This Approved Document has been written on the assumption that the building concerned will be properly managed.

Building Regulations do not impose any requirements on the management of a building. However, in developing an appropriate fire safety design for a building it may be necessary to consider the way in which it will be managed. A design which relies on an unrealistic or unsustainable management regime cannot be considered to have met the requirements of the Regulations.

Once the building is in use the management regime should be maintained and any variation in that regime should be the subject of a suitable risk assessment. Failure to take proper management responsibility may result in the prosecution of an employer, building owner or occupier under legislation such as the Regulatory Reform (Fire Safety) Order 2005.

PROPERTY PROTECTION

0.14 There are often many stakeholders, including insurers, who have a valid interest in the fire protection measures which are incorporated into a building's design. To ensure that the most effective fire protection measures are applied which are appropriate to the specific property, early consultation with the main stakeholders is essential. Failure to consult with stakeholders at an early stage could result in additional measures being required after completion, the use of the building being restricted, or insurance premiums and/or deductibles being increased.

Building Regulations are intended to ensure that a reasonable standard of life safety is provided, in case of fire. The protection of property, including the building itself, often requires additional measures and insurers will, in general, seek their own higher standards, before accepting the insurance risk.

Many insurers use the Fire Protection Association's (FPA) Design Guide for the fire protection of buildings as a basis for providing guidance to the building designer on what they require. Insurers' key objectives for achieving satisfactory standards of property protection are:

a. to limit damage to the fabric of the building caused by heat, smoke and firefighting water.

b. to limit damage to the contents of the building caused by heat, smoke and firefighting water.

c. to allow the business to be trading in as short a time as possible following a fire, thus limiting business interruption.

The FPA Design Guide is a suite of publications which incorporate:

a. An "Essential Principles" document which describes functional requirements.

b. A range of "Design Principles" documents which provide guidance for common building situations.

c. Separate "Core Documents" which expand upon guidance and explain construction detail which will deliver functional requirements.

Further information can be obtained from the FPA website: www.thefpa.co.uk.

Basic objectives for property protection

- To minimise the effect of fire on a business
- To limit the effects of business interruption
- To allow a business to be trading within 24 hours of a fire
- To protect the buildings within a business

These objectives for the protection of property can be addressed by meeting the principles set out in the *FPA Design guide: Essential principles* which are to be achieved in the design and construction of commercial and industrial premises.

The essential principles are:

- The building shall be constructed in such a manner that if a fire starts, the extent of fire and smoke damage will be minimised and confined as close to the source of fire outbreak as is practical/feasible;
- With the exception of joinery products, the building shall be constructed from building materials/products that will not make a significant contribution to the early stages of a fire or contribute to the spread of fire;

- Suitable measures will be taken for the prevention of premature structural collapse and excessive deflection;
- Consideration should be given at the design stage regarding potential damage from firefighting water and to ensure, as far as practical, that the effect on the environment of the fire effluent will be minimised;
- As a minimum, all fire protection products shall be third-party certified to an appropriate product or performance-based standard (attestation level 1 for CE marking);
- All fire protection products/systems shall be installed by adequately trained specialist installers;
- The building shall be fitted with an appropriate automatic fire alarm system;
- The fire protection systems shall be regularly maintained so that they are able to perform their intended function throughout the life of the building;
- There shall be adequate provision to prevent an arson attack;
- The building shall be so constructed that fire cannot spread into the premises from an adjoining building or other external fire source;
- The building owner shall ensure an adequate standard of fire safety management throughout the life of the building;
- Any fuel-burning appliance and services or electrical appliance and services shall be designed, constructed and installed in a manner that reduces their potential as an accidental source of ignition.

Guidance on property protection issues for schools is given in Building Bulletin (BB) 100 published by DfES. This gives advice on assessing the financial and social risk of school fires and advocates the use of fire suppression or additional compartmentation where the risk is justified.

From a property protection point of view, risk assessments on the majority of educational establishments tend to suggest that such measures are needed (see also the FPA sector guide, *Education Buildings*).

Guidance for asset protection in the Civil and Defence Estates is given in the Crown Fire Standards published by the Property Advisers to the Civil Estate (PACE).

INDEPENDENT SCHEMES OF CERTIFICATION AND ACCREDITATION

0.15 Since the performance of a system, product, component or structure is dependent upon satisfactory site installation, testing and maintenance, independent schemes of certification and accreditation of installers and maintenance firms of such will provide confidence in the appropriate standard of workmanship being provided.

Confidence that the required level of performance can be achieved will be demonstrated by the use of a system, material, product or structure which is provided under the arrangements of a product conformity certification scheme and an accreditation of installers scheme.

Third party accredited product conformity certification schemes not only provide a means of identifying materials and designs of systems, products or structures which have demonstrated that they have the requisite performance in fire, but additionally provide confidence that the systems, materials, products or structures actually supplied are provided to the same specification or design as that tested/assessed.

Third party accreditation of installers of systems, materials, products or structures provides a means of ensuring that installations have been conducted by knowledgeable contractors to appropriate standards, thereby increasing the reliability of the anticipated performance in fire.

Building Control Bodies may accept the certification of products, components, materials or structures under such schemes as evidence of compliance with the relevant standard. Similarly, Building Control Bodies may accept the certification of the installation or maintenance of products, components, materials or structures under such schemes as evidence of compliance with the relevant standard. Nonetheless, a Building Control Body will wish to establish, in advance of the work, that any such scheme is adequate for the purposes of the Building Regulations.

Many certification bodies which approve such schemes are accredited by UKAS.

SPRINKLER SYSTEMS

0.16 Sprinkler systems installed in buildings can reduce the risk to life and significantly reduce the degree of damage caused by fire. Sprinkler protection can also sometimes be used as a compensatory feature where the provisions of this Approved Document are varied in some way. Where sprinklers are provided, it is normal practice to provide sprinkler protection throughout a building. However, where the sprinklers are being installed as a compensatory feature to address a specific risk or hazard, it may be acceptable to protect only part of a building. Further guidance can also be found in *Sprinklers for Safety: Use and Benefits of Incorporating*

Sprinklers in Buildings and Structures, BAFSA 2006 (ISBN: 0 95526 280 1).

From a property protection perspective, all sprinkler installations should be in accordance with the *LPC Rules for automatic sprinkler installations*, published by the FPA. This incorporates BS EN 12845 but provides guidance in the form of technical bulletins to meet additional insurance requirements.

There are many alternative or innovative fire suppression systems available. Where these are used, it is necessary to ensure that such systems have been designed and tested for use in buildings and are fit for their intended purpose. **Advice from insurers should be obtained before specifying such systems.**

0.17 Where a sprinkler system is specifically recommended within this document it should be provided throughout the building or separated part and be designed and installed in accordance with either:

a. for dwellings and residential buildings, BS 9251:2005 *Sprinkler systems for residential and domestic occupancies – Code of practice* and BS DD 252 *Components for residential sprinkler systems – Specification and test methods for residential sprinklers*; or

b. for non-residential buildings or dwellings and residential buildings outside the scope of BS 9251, either:

   i. the requirements of BS 5306-2:1990, including the relevant hazard classification together with the additional requirements for life safety; or

   ii. the requirements of BS EN 12845:2004, including the relevant hazard classification together with the special requirements for life safety systems.

Note: Any sprinkler system installed to satisfy the requirements of Part B of the Building Regulations should be regarded as a life safety system. However, there may be some circumstances where a particular life safety requirement, specified in BS 5306-2 or BS EN 12845 is inappropriate or unnecessary.

0.18 Water supplies for non-residential sprinkler systems should consist of either:

a. for systems designed and installed to BS 5306-2:

   i. two single water supplies complying with BS 5306-2, clause 13.1.2 where each is independent of the other; or

   ii. two stored water supplies, where:

      1. gravity or suction tanks should be either Type A, Type D or their equivalent, (see BS 5306-2 clause 17.4.11.6); and

      2. any pump arrangements should comply with BS 5306-2 clause 17.4.1.5; and

      3. the capacity of each tank should be equivalent to at least half the specified minimum water volume of a single full capacity tank, appropriate to the hazard; or

      4. one tank should be equivalent to half the specified water volume of a single full capacity tank and the other shall not be less than half the minimum volume of a reduced capacity tank (see BS 5306-2, Table 25), appropriate to the hazard; and

Note: The requirements for inflow should be met.

      5. whichever water storage arrangement is used at (3) or (4) above, the total design capacity of the water supply, including any inflow for a reduced capacity tank should be at least equivalent to a single full holding capacity tank complying with Table 21, 22, 23 or 24, as appropriate to the hazard and pipework design.

b. for systems designed and installed to BS EN 12845:

   i. two single water supplies complying with BS EN 12845, clause 9.6.1 where each is independent of the other; or

   ii. two stored water supplies, where:

      1. gravity or suction tanks should satisfy all the requirements of BS EN 12845 clause 9.6.2 b) other than capacity; and

      2. any pump arrangements should comply with BS EN 12845 clause 10.2 and

      3. the capacity of each tank is equivalent to half the specified minimum water volume of a single full capacity tank, appropriate to the hazard; or

      4. one tank should be at least equivalent to half the specified water volume of a single full capacity tank and the other shall not be less than the minimum volume of a reduced capacity tank BS EN12845 clause 9.3.4, appropriate to the hazard; and

Note: The requirement for inflow should be met.

      5. whichever water storage arrangement is used at (3) or (4) above, the total capacity of the water supply, including any inflow for a reduced capacity tank should be at least equivalent to a single full holding capacity tank complying with BS EN12845, Table 9, 10 or clause 9.3.2.3 as appropriate to the hazard and pipework design.

Where pumps are used to draw water from two tanks, then each pump should be arranged to draw water from either tank and arranged so that any one pump or either tank could be isolated.

The sprinkler water supplies should generally not be used as connections for other services or other fixed firefighting systems.

INCLUSIVE DESIGN

0.19 The fire safety aspects of the Building Regulations are made for securing reasonable standards of health and safety of persons in and about buildings. This is intended to include all people, including people with disabilities.

Part M of the Regulations, *Access to and use of buildings,* requires reasonable provision for access by people to buildings. Regardless of compliance with Building Regulations, there will also be obligations under the Disability Discrimination Act 1995 for service providers and employers to consider barriers created by physical features in buildings.

People, regardless of disability, age or gender, should be able to gain access to buildings and use their facilities, both as visitors and as people who live or work in them.

As such the fire safety measures incorporated into a building will need to take account of the needs of all those persons who may have access to the building. It is not appropriate, except in exceptional circumstances, to presume that certain groups of people will be excluded from a building because of its use.

The provisions set out in this Approved Document are considered to be a reasonable standard for most buildings. However, there may be some people whose specific needs are not addressed. In some situations additional measures may be needed to accommodate these needs. This should be done on a case by case basis.

MATERIAL ALTERATION

0.20 Under Regulation 3, the term "material alteration" is defined by reference to a list of "relevant requirements" of Schedule 1 to the Building Regulations. That list includes the requirements of Parts B1, B3, B4 and B5. This means that an alteration which, at any stage of the work, results in a building being less satisfactory than it was before in relation to compliance with the requirements of Parts B1, B3, B4 or B5 is a material alteration, and is therefore controlled by Regulation 4 as it is classed as "building work". Regulation 4(1) requires that any building work carried out in relation to a material alteration complies with the applicable requirements of Schedule 1 to the Regulations, while Regulation 4(2) requires that once that building work has been completed, the building as a whole must comply with the relevant requirements of Schedule 1 or, where it did not comply before, must be no more unsatisfactory than it was before the work was carried out.

ALTERNATIVE APPROACHES

0.21 The fire safety requirements of the Building Regulations should be satisfied by following the relevant guidance given in this Approved Document. However, Approved Documents are intended to provide guidance for some of the more common building situations and there may well be alternative ways of achieving compliance with the requirements.

If other codes or guides are adopted, the relevant recommendations concerning fire safety in the particular publication should be followed, rather than a mixture of the publication and provisions in the relevant sections of this Approved Document. However, there may be circumstances where it is necessary to use one publication to supplement another.

Guidance documents intended specifically for assessing fire safety in **existing buildings** will often include provisions which are less onerous than those set out on this Approved Document or other standards applicable to new buildings. As such, these documents are unlikely to be appropriate for use where building work, controlled by the Regulations, is proposed.

Note: Buildings for some particular industrial and commercial activities presenting a special fire hazard, e.g. those involved with the sale of fuels, may require additional fire precautions to those detailed in this Approved Document.

FPA Design guide sector guides are available to cover specific types of buildings that require particular attention.

British Standards

0.22 Compliance with a British Standard does not of itself confer immunity from legal obligations. British Standards can, however, provide a useful source of information which could be used to supplement or provide an alternative to the guidance given in this Approved Document.

When an Approved Document makes reference to a named standard, the relevant version of the standard is the one listed at the end of the publication. However, if this version of the standard has been revised or updated by the issuing standards body, the new version may be used as a source of guidance provided it continues to address the relevant requirements of the Regulations.

Drafts for Development (DDs) are not British Standards. They are issued in the DD series of publications and are of a provisional nature. They are intended to be applied on a provisional basis so that information and experience of their practical application may be obtained and the document developed. Where the recommendations of a DD are adopted then care should be taken to ensure that the requirements of the Building Regulations are adequately met. Any observations that a user may have in relation to any aspect of a DD should be passed on to BSI.

Health care premises

0.23 Health care premises are quite diverse and can be used by a variety of patients, often requiring different types of care to suit their specific needs. The choice of fire safety strategy is dependent upon the way a building is designed, furnished, staffed and managed and the level of dependency of the patients.

In parts of health care premises designed to be used by patients where there are people who are bedridden or who have very restricted mobility, the principle of total evacuation of a building in the event of fire may be inappropriate. It is also unrealistic to suppose that all patients will leave without assistance. In this and other ways the specialised nature of some health care premises demands a different approach to the provision of means of escape, from much of that embodied by the guidance in this Approved Document.

The Department of Health has prepared a set of guidance documents on fire precautions in health care buildings, under the general title of *'Firecode'*, taking into account the particular characteristics of these buildings. These documents may also be used for non-NHS health care premises.

The design of fire safety in health care premises is covered by Health Technical Memorandum (HTM) 05-02 *Guidance in support of functional provisions for healthcare premises*. Part B of the Building Regulations will typically be satisfied where the guidance in that document is followed. Where work to existing healthcare premises is concerned the guidance in the appropriate section of the relevant *Firecode* should be followed.

Unsupervised group homes

0.24 Where an existing house of one or two storeys is to be put to use as an unsupervised group home for not more than six mental health service users, it should be regarded as a Purpose Group 1(c) building if the means of escape are provided in accordance with HTM 88: *Guide to fire precautions in NHS housing in the community for mentally handicapped (or mentally ill) people.* Where the building is new, it may be more appropriate to regard it as being in Purpose Group 2(b).

Note: *Firecode* contains managerial and other fire safety provisions which are outside the scope of Building Regulations.

Shopping complexes

0.25 Although the guidance in this Approved Document may be readily applied to individual shops, shopping complexes present a different set of escape problems. The design of units within a shopping complex should be compatible with the fire strategy for the complex as a whole. A suitable approach is given in BS 5588-10:1991.

Note: BS 5588-10:1991 applies more restrictive provisions to units with only one exit in covered shopping complexes.

Assembly buildings

0.26 There are particular problems that arise when people are limited in their ability to escape by fixed seating. This may occur at sports events, theatres, lecture halls and conference centres etc. Guidance on this and other aspects of means of escape in assembly buildings is given in Sections 3 and 5 of BS 5588-6:1991 and the relevant recommendations concerning means of escape in case of fire of that code should be followed in appropriate cases. In the case of buildings to which the Safety of Sports Grounds Act 1975 applies, the *Guide to safety at sports grounds* TSO (ISBN: 0 11341 001 8) should also be followed.

Schools

0.27 The design of fire safety in schools is covered by Building Bulletin (BB) 100 published by the DfES. Part B of the Building Regulations will typically be satisfied where the life safety guidance in that document is followed.

Buildings containing one or more atria

0.28 A building containing an atrium passing through compartment floors may need special fire safety measures. Guidance on suitable fire safety measures in these circumstances is to be found in BS 5588-7:1997 (see also paragraph 8.8). For shopping complexes see paragraph 0.25.

Sheltered housing

0.29 Whilst many of the provisions in this Approved Document for means of escape from flats are applicable to sheltered housing, the nature of the occupancy may necessitate some additional fire protection measures. The extent will depend on the form of the development. For example, a group of specially adapted bungalows or two-storey flats, with few communal facilities, need not be treated differently from other one or two-storey dwellinghouses or flats.

Fire safety engineering

0.30 Fire safety engineering can provide an alternative approach to fire safety. It may be the only practical way to achieve a satisfactory standard of fire safety in some large and complex buildings and in buildings containing different uses, e.g. airport terminals. Fire safety engineering may also be suitable for solving a problem with an aspect of the building design which otherwise follows the provisions in this document.

0.31 British Standard BS 7974 *Fire safety engineering in buildings* and supporting published documents (PDs) provide a framework and guidance on the design and assessment of fire safety measures in buildings. Following the discipline of BS 7974 should enable designers and Building Control Bodies to be aware of the relevant issues, the need to consider the complete fire-safety system and to follow a disciplined analytical framework.

0.32 Factors that should be taken into account include:

a. the anticipated probability of a fire occurring;

b. the anticipated fire severity;

c. the ability of a structure to resist the spread of fire and smoke;

d. the consequential danger to people in and around the building; and

e. **the anticipated property damage to the building and adjoining buildings, together with the associated business interruption.**

0.33 A wide variety of measures could be considered and incorporated to a greater or lesser extent, as appropriate in the circumstances. These include:

a. the adequacy of means to prevent fire;

b. early fire warning by an automatic detection and warning system;

c. the standard of means of escape;

d. provision of smoke control;

e. control of the rate of growth of a fire;

f. structural robustness and the adequacy of the structure to resist the effects of a fire;

g. the degree of fire containment;

h. fire separation between buildings or parts of buildings;

i. the standard of active measures for fire extinguishment or control;

j. facilities to assist the fire and rescue service;

k. availability of powers to require staff training in fire safety and fire routines;

l. consideration of the availability of any continuing control under other legislation that could ensure continued maintenance of such systems; and

m. management.

0.34 It is possible to use quantitative techniques to evaluate risk and hazard. Some factors in the measures listed above can be given numerical values in some circumstances. The assumptions made when quantitative methods are used need careful assessment.

Buildings of special architectural of historic interest

0.35 Some variation of the provisions set out in this document may also be appropriate where Part B applies to existing buildings, particularly in buildings of special architectural or historic interest, where adherence to the guidance in this document might prove unduly restrictive. In such cases it would be appropriate to take into account a range of fire safety features, some of which are dealt with in this document and some of which are not addressed in any detail and to set these against an assessment of the hazard and risk peculiar to the particular case.

The Requirement

This Approved Document deals with the following requirement from Part B of Schedule 1 to the Building Regulations 2000 (as amended).

Requirement	*Limits on application*
Means of warning and escape	
B1. The building shall be designed and constructed so that there are appropriate provisions for the early warning of fire, and appropriate means of escape in case of fire from the building to a place of safety outside the building capable of being safely and effectively used at all material times.	Requirement B1 does not apply to any prison provided under Section 33 of the Prison Act 1952 (power to provide prisons, etc.).

Insurers' Essential Principles

The relevant essential principles for property protection are:

- The building shall be fitted with an appropriate automatic fire alarm system;
- The fire protection systems shall be regularly maintained so that they are able to perform their intended function throughout the life of the building;
- The building owner shall ensure an adequate standard of fire safety management throughout the life of the building.

Guidance

Performance

In the Secretary of State's view the Requirement of B1 will be met if:

a. there are routes of sufficient number and capacity, which are suitably located to enable persons to escape to a place of safety in the event of fire;

b. the routes are sufficiently protected from the effects of fire where necessary;

c. the routes are adequately lit;

d. the exits are suitably signed; and

e. there are appropriate facilities to either limit the ingress of smoke to the escape route(s) or to restrict the fire and remove smoke;

f. all to an extent necessary that is dependent on the use of the building, its size and height; and

g. there is sufficient means for giving early warning of fire for persons in the building.

Introduction

B1.i These provisions relate to building work and material changes of use which are subject to the functional requirement B1 and they may therefore affect new or existing buildings. They are concerned with the measures necessary to ensure reasonable facilities for means of escape in case of fire. They are only concerned with structural fire precautions where these are necessary to safeguard escape routes.

They assume that, in the design of the building, reliance should not be placed on external rescue by the Fire and Rescue Service nor should it be based on a presumption that the Fire and Rescue Service will attend an incident within a given time. This Approved Document has been prepared on the basis that, in an emergency, the occupants of any part of a building should be able to escape safely without any external assistance.

Special considerations, however, apply to some institutional buildings in which the principle of evacuation with assistance from staff is necessary.

Analysis of the problem

B1.ii The design of means of escape and the provision of other fire safety measures such as a fire alarm system (where appropriate), should be based on an assessment of the risk to the occupants should a fire occur. The assessment should take into account the nature of the building structure, the use of the building, the processes undertaken and/or materials stored in the building; the potential sources of fire; the potential of fire spread through the building; and the standard of fire safety management proposed. Where it is not possible to identify with any certainty any of these elements, a judgement as to the likely level of provision must be made.

B1.iii Fires do not normally start in two different places in a building at the same time. Initially a fire will create a hazard only in the part in which it starts and it is unlikely, at this stage, to involve a large area. The fire may subsequently spread to other parts of the building, usually along the circulation routes. The items that are the first to be ignited are often furnishings and other items not controlled by the regulations. It is less likely that the fire will originate in the structure of the building itself and the risk of it originating accidentally in circulation areas, such as corridors, lobbies or stairways, is limited, provided that the combustible content of such areas is restricted.

B1.iv The primary danger associated with fire in its early stages is not flame but the smoke and noxious gases produced by the fire. They cause most of the casualties and may also obscure the way to escape routes and exits. Measures designed to provide safe means of escape must therefore provide appropriate arrangements to limit the rapid spread of smoke and fumes.

Criteria for means of escape

B1.v The basic principles for the design of means of escape are:

a. that there should be alternative means of escape from most situations; and

b. where direct escape to a place of safety is not possible, it should be possible to reach a place of relative safety, such as a protected stairway, which is on a route to an exit, within a reasonable travel distance. In such cases the means of escape will consist of two parts, the first being unprotected in accommodation and circulation areas and the second in protected stairways (and in some circumstances protected corridors).

Note: Some people, for example those who use wheelchairs, may not be able to use stairways without assistance. For them evacuation involving the use of refuges on escape routes and either assistance down (or up) stairways or the use of suitable lifts will be necessary.

The ultimate place of safety is the open air clear of the effects of the fire. However, in modern buildings which are large and complex, reasonable safety may be reached within the building, provided suitable planning and protection measures are incorporated.

B1.vi For the purposes of the Building Regulations, the following are not acceptable as means of escape:

a. lifts (except for a suitably designed and installed evacuation lift – see paragraph 5.39);

b. portable ladders and throw-out ladders; and

c. manipulative apparatus and appliances: e.g. fold-down ladders and chutes.

Escalators should not be counted as providing predictable exit capacity, although it is recognised that they are likely to be used by people who are escaping. Mechanised walkways could be accepted and their capacity assessed on the basis of their use as a walking route, while in the static mode.

Alternative means of escape

B1.vii There is always the possibility of the path of a single escape route being rendered impassable by fire, smoke or fumes. Ideally, therefore people should be able to turn their backs on a fire wherever it occurs and travel away from it to a final exit or protected escape route leading to a place of safety. However, in certain conditions a single direction of escape (a dead end) can be accepted as providing reasonable safety. These conditions depend on the use of the building and its associated fire risk, the size and height of the building, the extent of the dead end and the numbers of persons accommodated within the dead end.

Unprotected and protected escape routes

B1.viii The unprotected part of an escape route is that part which a person has to traverse before reaching either the safety of a final exit or the comparative safety of a protected escape route, i.e. a protected corridor or protected stairway.

Unprotected escape routes should be limited in extent so that people do not have to travel excessive distances while exposed to the immediate danger of fire and smoke.

Even with protected horizontal escape routes, the distance to a final exit or protected stairway needs to be limited because the structure does not give protection indefinitely.

B1.ix Protected stairways are designed to provide virtually 'fire sterile' areas which lead to places of safety outside the building. Once inside a protected stairway, a person can be considered to be safe from immediate danger from flame and smoke. They can then proceed to a place of safety at their own pace. To enable this to be done, flames, smoke and gases must be excluded from these escape routes, as far as is reasonably possible, by fire-resisting structures or by an appropriate smoke control system, or by a combination of both these methods. This does not preclude the use of unprotected stairs for day-to-day circulation, but they can only play a very limited role in terms of means of escape due to their vulnerability in fire situations.

Security

B1.x The need for easy and rapid evacuation of a building in case of fire may conflict with the control of entry and exit in the interest of security. Measures intended to prevent unauthorised access can also hinder entry of the fire and rescue service to rescue people trapped by fire.

Potential conflicts should be identified and resolved at the design stage and not left to ad hoc expedients after completion. The architectural liaison officers attached to most police forces are a valuable source of advice. Some more detailed guidance on door security in buildings is given in paragraphs 5.11 and 5.12.

Use of the document

B1.xi Section 1 deals with fire alarm and fire detection systems in all buildings. Section 2 deals with means of escape from blocks of flats and Sections 3 and 4 with buildings other than flats. Section 3 concerns the design of means of escape on one level (the horizontal phase in multi-storey buildings). Section 4 deals with stairways and the vertical phase of the escape route. Section 5 gives guidance on matters common to all parts of the means of escape.

Section 1: Fire alarm and fire detection systems

Introduction

1.1 Provisions are made in this section for suitable arrangements to be made in all buildings to give early warning in the event of fire.

Paragraphs 1.2 to 1.23 deal with flats and paragraphs 1.24 to 1.37 with buildings other than flats. Paragraph 1.38 is applicable to all uses.

Flats

1.2 Provisions are made in this section for suitable arrangements to be made in flats to give early warning in the event of fire.

General

1.3 In most flats, the installation of smoke alarms or automatic fire detection and alarm systems, can significantly increase the level of safety by automatically giving an early warning of fire. The following guidance is appropriate for most flats. However, where it is known that the occupants of a proposed flat are at a special risk from fire, it may be more appropriate to provide a higher standard of protection (i.e. additional alarms).

1.4 All new flats should be provided with a fire detection and fire alarm system in accordance with the relevant recommendations of BS 5839-6:2004 *Code of practice for the design, installation and maintenance of fire detection and fire alarm systems in dwellings* to at least a Grade D Category LD3 standard.

1.5 The smoke and heat alarms should be mains-operated and conform to BS 5446-1:2000 or BS 5446-2:2003 respectively: *Fire detection and fire alarm devices for dwellings*, Part 1 *Specification for smoke alarms;* or Part 2 *Specification for heat alarms*. They should have a standby power supply such as a battery (either rechargeable or non-rechargeable) or capacitor. More information on power supplies is given in clause 15 of BS 5839-6.

Note: BS 5446-1 covers smoke alarms based on ionization chamber smoke detectors and optical (photo-electric) smoke detectors. The different types of detector respond differently to smouldering and fast-flaming fires. Either type of detector is generally suitable. However, the choice of detector type should, if possible, take into account the type of fire that might be expected and the need to avoid false alarms. Optical detectors tend to be less affected by low levels of 'invisible' particles, such as fumes from kitchens, that often cause false alarms. Accordingly, they are generally more suitable than ionization chamber detectors for installation in circulation spaces adjacent to kitchens.

Material alterations

1.6 Where new habitable rooms are provided above the ground floor level, or where they are provided at ground floor level and there is no final exit from the new room, a fire detection and fire alarm system should be installed. Smoke alarms should be provided in the circulation spaces of the dwelling in accordance with paragraphs 1.10 to 1.18 to ensure that any occupants of the new rooms are warned of any fire that may impede their escape.

Sheltered housing

1.7 The detection equipment in a sheltered housing scheme with a warden or supervisor should have a connection to a central monitoring point (or alarm receiving centre) so that the person in charge is aware that a fire has been detected in one of the flats and can identify the flat concerned. These provisions are not intended to be applied to the common parts of a sheltered housing development, such as communal lounges, or to sheltered accommodation in the Institutional or Other residential purpose groups. Means of warning in such facilities should be considered on a case by case basis following the general guidance for buildings other than flats given in paragraphs 1.24 to 1.38.

Student accommodation

1.8 Some student residential accommodation is constructed in the same way as a block of flats. Where groups of up to six students share a self-contained flat with its own entrance door, constructed on the compartmentation principles for flats in Section 7 (B3), it is appropriate to provide a separate automatic detection system within each flat. Where a general evacuation is required (e.g. halls of residence), the alarm system should follow the guidance for buildings other than flats given in paragraphs 1.24. to 1.38.

Positioning of smoke and heat alarms

1.9 Detailed guidance on the design and installation of fire detection and alarm systems in flats is given in BS 5839-6. However, the following guidance is appropriate to most common situations.

1.10 Smoke alarms should normally be positioned in the circulation spaces between sleeping spaces and places where fires are most likely to start (e.g. kitchens and living rooms) to pick up smoke in the early stages.

1.11 There should be at least one smoke alarm on every storey of a flat.

1.12 Where the kitchen area is not separated from the stairway or circulation space by a door, there should be a compatible interlinked heat detector or heat alarm in the kitchen, in addition to whatever smoke alarms are needed in the circulation space(s);

1.13 Where more than one alarm is installed they should be linked so that the detection of smoke by one unit operates the alarm signal in all of them. The manufacturers' instructions about the maximum number of units that can be linked should be observed.

1.14 Smoke alarms/detectors should be sited so that:

a. there is a smoke alarm in the circulation space within 7.5m of the door to every habitable room;

b. they are ceiling-mounted and at least 300mm from walls and light fittings (unless in the case of light fittings there is test evidence to prove that the proximity of the light fitting will not adversely affect the efficiency of the detector). Units designed for wall-mounting may also be used provided that the units are above the level of doorways opening into the space and they are fixed in accordance with manufacturers' instructions; and

c. the sensor in ceiling-mounted devices is between 25mm and 600mm below the ceiling (25-150mm in the case of heat detectors or heat alarms).

Note: This guidance applies to ceilings that are predominantly flat and horizontal.

1.15 It should be possible to reach the smoke alarms to carry out routine maintenance, such as testing and cleaning, easily and safely. For this reason smoke alarms should not be fixed over a stair or any other opening between floors.

1.16 Smoke alarms should not be fixed next to or directly above heaters or air-conditioning outlets. They should not be fixed in bathrooms, showers, cooking areas or garages, or any other place where steam, condensation or fumes could give false alarms.

1.17 Smoke alarms should not be fitted in places that get very hot (such as a boiler room), or very cold (such as an unheated porch). They should not be fixed to surfaces which are normally much warmer or colder than the rest of the space, because the temperature difference might create air currents which move smoke away from the unit.

1.18 A requirement for maintenance can not be made as a condition of passing plans by the Building Control Body. However, the attention of developers and builders is drawn to the importance of providing the occupants with information on the use of the equipment and on its maintenance (or guidance on suitable maintenance contractors). See paragraph 0.13.

Note: BS 5839-1 and BS 5839-6 recommend that occupiers should receive the manufacturers' instructions concerning the operation and maintenance of the alarm system.

Power supplies

1.19 The power supply for a smoke alarm system should be derived from the flat's mains electricity supply. The mains supply to the smoke alarm(s) should comprise a single independent circuit at the flat's main distribution board (consumer unit) or a single regularly used local lighting circuit. This has the advantage that the circuit is unlikely to be disconnected for any prolonged period. There should be a means of Isolating power to the smoke alarms without isolating the lighting.

1.20 The electrical installation should comply with Approved Document P (Electrical safety).

1.21 Any cable suitable for domestic wiring may be used for the power supply and interconnection to smoke alarm systems. It does not need any particular fire survival properties. Any conductors used for interconnecting alarms (signalling) should be readily distinguishable from those supplying mains power, e.g. by colour coding.

Note: Mains powered smoke alarms may be interconnected using radio-links, provided that this does not reduce the lifetime or duration of any standby power supply below 72 hours. In this case, the smoke alarms may be connected to separate power circuits (see paragraph 1.19)

1.22 Other effective options exist and are described in BS 5839: Parts 1 and 6. For example, the mains supply may be reduced to extra low voltage in a control unit incorporating a standby trickle-charged battery, before being distributed at that voltage to the alarms.

Design and installation of systems

1.23 It is essential that fire detection and fire alarm systems are properly designed, installed and maintained. Where a fire alarm system is installed, an installation and commissioning certificate should be provided. Third party certification schemes for fire protection products and related services are an effective means of providing the fullest possible assurances, offering a level of quality, reliability and safety.

Buildings other than flats

General

1.24 To select the appropriate type of fire alarm/detection system that should be installed into a particular building, the type of occupancy and means of escape strategy (e.g. simultaneous, phased or progressive horizontal evacuation) must be determined.

1.25 For example, if occupants normally sleep on the premises e.g. residential accommodation, the threat posed by a fire is much greater than that in premises where the occupants are normally alert. Where the means of escape is based on simultaneous evacuation, operation of a manual call point or fire detector should give an almost instantaneous warning from all the fire alarm sounders. However, where the means of escape is based on phased evacuation, then a staged alarm system is appropriate. Such a system enables two or more stages of alarm to be given within a particular area, e.g. "alert" or "evacuate" signals.

Note: the term fire detection system is used here to describe any type of automatic sensor network and associated control and indicating equipment. Sensors may be sensitive to smoke, heat, gaseous combustion products or radiation. Normally the control and indicating equipment operates a fire alarm system and it may perform other signalling or control functions as well. Automatic sprinkler systems can also be used to operate a fire alarm system.

1.26 The factors which have to be considered when assessing what standard of fire alarm or automatic fire detection system is to be provided will vary widely from one set of premises to another. Therefore the appropriate standard will need to be considered on a case by case basis.

Note: General guidance on the standard of automatic fire detection that **may** need to be provided within a building can be found in Table A1 of BS 5839-1:2002.

Fire alarm systems

1.27 All buildings should have arrangements for detecting fire. In most buildings fires are detected by people, either through observation or smell and therefore often nothing more will be needed.

1.28 In small buildings/premises the means of raising the alarm may be simple. For instance, where all occupants are near to each other a shouted warning "FIRE" by the person discovering the fire may be all that is needed. In assessing the situation, it must be determined that the warning can be heard and understood throughout the premises, including for example the toilet areas. In other circumstances, manually operated sounders (such as rotary gongs or handbells) may be used. Alternatively a simple manual call point combined with a bell, battery and charger may be suitable.

1.29 In all other cases, the building should be provided with a suitable electrically operated fire warning system with manual call points sited adjacent to exit doors and sufficient sounders to be clearly audible throughout the building.

1.30 An electrically operated fire alarm system should comply with BS 5839-1:2002 *Fire detection and alarm systems for buildings, Code of practice for system design, installation commissioning and maintenance.*

BS 5839-1 specifies three categories of system, i.e. category L for the protection of life; category M manual alarm systems; category P for property protection. Category L systems are sub-divided into:

L1 – systems installed throughout the protected building;

L2 – systems installed only in defined parts of the protected building (a category L2 system should normally include the coverage required of a category L3 system);

L3 – systems designed to give a warning of fire at an early enough stage to enable all occupants, other than possibly those in the room of fire origin, to escape safely, before the escape routes are impassable owing to the presence of fire, smoke or toxic gases;

L4 – systems installed within those parts of the escape routes comprising circulation areas and circulation spaces, such as corridors and stairways; and

L5 – systems in which the protected area(s) and/or the location of detectors is designed to satisfy a specific fire safety objective (other than that of a category L1, L2, L3 or L4 system).

The findings of the fire risk assessment may, for the purpose of property protection and business resilience, lead to the requirement for enhanced property protection. Hence consultation should always be made with insurers.

For larger/complex risks, particularly industrial/commercial risks, the insurance requirement may be for minimum L1/L2 standards i.e. L3 – L5 are partial-only systems and not fully comprehensive systems.

Type P systems are sub-divided into P1 – systems installed throughout the protected building and P2 – systems installed only in defined parts of the protected building.

1.31 Call points for electrical alarm systems should comply with BS 5839-2:1983, or Type A of BS EN 54-11:2001 and these should be installed in accordance with BS 5839-1. Type B call points should only be used with the approval of the Building Control Body.

BS EN 54-11 covers two types of call points, Type A (direct operation) in which the change to the alarm condition is automatic (i.e. without the need for further manual action) when the frangible element is broken or displaced; and Type B (indirect operation) in which the change to the alarm condition requires a separate manual operation of the operating element by the user after the frangible element is broken or displaced.

1.32 If it is considered that people might not respond quickly to a fire warning, or where people are unfamiliar with the fire warning arrangements, consideration may be given to installing a voice alarm system. Such a system could form part of a public address system and give both an audible signal and verbal instructions in the event of fire.

The fire warning signal should be distinct from other signals which may be in general use and be accompanied by clear verbal instructions.

If a voice alarm system is to be installed, it should comply with BS 5839-8:1998 *Code of practice for the design, installation and servicing of voice alarm systems.*

1.33 In certain premises, e.g. large shops and places of assembly, an initial general alarm may be undesirable because of the number of members of the public present. The need for fully trained staff to effect pre-planned procedures for safe evacuation will therefore be essential. Actuation of the fire alarm system will cause staff to be alerted, e.g. by discreet sounders, personal paging systems etc. Provision will normally be made for full evacuation of the premises by sounders or a message broadcast over the public address system. In all other respects, any staff alarm system should comply with BS 5839-1.

Warnings for people with impaired hearing

1.34 A suitable method of warning (e.g. a visual and audible fire alarm signal) should be provided in buildings where it is anticipated that one or more persons with impaired hearing may be in relative isolation (e.g. hotel bedrooms and sanitary accomodation) and where there is no other suitable method of alerting them.

In buildings such as schools, colleges and offices where the population is controlled, a vibrating paging system may be more appropriate. This could also be used for alerting people with other disabilities.

Clause 18 of BS 5839-1:2002 provides detailed guidance on the design and selection of fire alarm warnings for people with impaired hearing.

Automatic fire detection and fire alarm systems

1.35 Automatic fire detection and alarms in accordance with BS 5839-1 should be provided in Institutional and Other residential occupancies.

1.36 Automatic fire detection systems are not normally needed in non-residential occupancies. However, there are often circumstances where a fire detection system in accordance with BS 5839-1 may be needed. For example:

a. to compensate for some departure from the guidance elsewhere in this document;

b. as part of the operating system for some fire protection systems, such as pressure differential systems or automatic door releases;

c. where a fire could break out in an unoccupied part of the premises (e.g. a storage area or basement that is not visited on a regular basis, or a part of the building that has been temporarily vacated) and prejudice the means of escape from any occupied part(s) of the premises.

Note 1: Guidance on the provision of automatic fire detection within a building which is designed for phased evacuation can be found in paragraph 4.29.

Note 2: Where an atrium building is designed in accordance BS 5588-7:1997, then the relevant recommendations in that code for the installation of fire alarm/fire detection systems for the design option(s) selected should be followed.

Design and installation of systems

1.37 It is essential that fire detection and fire warning systems are properly designed, installed and maintained. Where a fire alarm system is installed, an installation and commissioning certificate should be provided. Third party certification schemes for fire protection products and related services are an effective means of providing the fullest possible assurances, offering a level of quality, reliability and safety (see paragraph 0.15).

Interface between fire detection and fire alarm systems and other systems

1.38 Fire detection and fire alarm systems are sometimes used to initiate the operation, or change of state, of other systems, such as smoke control systems, fire extinguishing systems, release arrangements for electrically held-open fire doors and electronically locked exit doors. It is essential that the interface between the fire detection and fire alarm system and any other system required for compliance with the Building Regulations is designed to achieve a high degree of reliability. Particular care should be taken if the interface is facilitated via another system, such as an access control system. Where any part of BS 7273 applies to actuation of other systems, the recommendations of that standard should be followed.

Property protection

1.39 For the purpose of property protection and business resilience, in unsprinklered buildings, a fire detection system should be provided which complies with BS 5839-1:2002 +A2:2008 and conforms to design of Category P1/M (property risk protection). Only equipment approved by a body acceptable to insurers should be specified.

This class of system would provide detection throughout the building in addition to manual call points.

In buildings provided with sprinklers*, a fire detection system should be provided which complies with BS 5839-1:2002 +A2:2008 and conforms to design of Category P2/M (property risk protection). In specific risk situations, insurers may require Category P1/M (property risk protection).

This class of system would provide early warning of fires in areas of high fire hazard level, or areas in which the risk to property or business continuity from fire is high.

A monitored remote connection to a central alarm receiving centre must be provided for both types of system to enable swift attendance by the fire and rescue service outside normal working hours.

*'Sprinklered' means that the building is fitted throughout with an automatic sprinkler system in accordance with paragraph 0.16.

Means of escape from flats

Introduction

2.1 The means of escape from a flat with a floor not more than 4.5m above ground level is relatively simple to provide. Few provisions are specified in this document beyond ensuring that means are provided for giving early warning in the event of fire (see Section 1) and that suitable means are provided for emergency egress from these storeys.

With increasing height more complex provisions are needed because emergency egress through upper windows becomes increasingly hazardous.

2.2 The guidance in this section deals with some common arrangements of flat design. Other, less common, arrangements (for example flats entered above or below accommodation level, or flats containing galleries) are acceptable. Guidance on these is given in clauses 9 and 10 of BS 5588-1:1990.

2.3 The provisions for means of escape for flats are based on the assumption that:

a. the fire is generally in a flat;

b. there is no reliance on external rescue (e.g. by a portable ladder);

c. measures in Section 8 (B3) provide a high degree of compartmentation and therefore a low probability of fire spread beyond the flat of origin, so that simultaneous evacuation of the building is unlikely to be necessary; and

d. although fires may occur in the common parts of the building, the materials and construction used there should prevent the fabric from being involved beyond the immediate vicinity (although in some cases communal facilities exist which require additional measures to be taken).

2.4 There are two distinct components to planning means of escape from buildings containing flats; escape from within each flat and escape from each flat to the final exit from the building.

Paragraphs 2.5 to 2.18 deal with the means of escape within each unit, i.e. within the private domestic area. Paragraphs 2.19 to 2.48 deal with the means of escape in the common areas of the building. Guidance on mixed use buildings is given in paragraphs 2.50 to 2.51 and for live/work units in 2.52.

General provisions

Inner rooms

2.5 A room whose only escape route is through another room is at risk if a fire starts in that other room.

This situation may arise with open-plan layouts and galleries. Such an arrangement is only acceptable where the inner room is:

a. a kitchen;

b. a laundry or utility room;

c. a dressing room;

d. a bathroom, WC, or shower room;

e. any other room on a floor not more than 4.5m above ground level which complies with paragraph 2.6, 2.10, or 2.11b as appropriate; or

f. a gallery which complies with paragraph 2.8.

Note: A room accessed only via an inner room (an inner-inner room) may be acceptable if it complies with the above, not more than one door separates the room from an interlinked smoke alarm and none of the access rooms are a kitchen.

Basements

2.6 Because of the risk that a single stairway may be blocked by smoke from a fire in the basement or ground storey, if the basement storey contains any habitable room, either provide:

a. an external door or window suitable for egress from the basement (see paragraph 2.9); or

b. a protected stairway leading from the basement to a final exit.

Balconies and flat roofs

2.7 Any balcony outside an alternative exit to a flat more than 4.5m above ground level should be a common balcony and meet the conditions in paragraph 2.17.

A flat roof forming part of a means of escape should comply with the following provisions:

a. the roof should be part of the same building from which escape is being made;

b. the route across the roof should lead to a storey exit or external escape route; and

c. the part of the roof forming the escape route and its supporting structure, together with any opening within 3m of the escape route, should provide 30 minutes fire resistance (see Appendix A Table A1).

Note: Where a balcony or flat roof is provided for escape purposes, guarding may be needed, in which case it should meet the provisions in Approved Document K *Protection from falling, collision and impact.*

Galleries

2.8 A gallery should be provided with an alternative exit; or, where the gallery floor is not more than 4.5m above ground level, an emergency egress window which complies with paragraph 2.9. Where the gallery floor is not provided with an alternative exit or escape window, it should comply with the following:

a. the gallery should overlook at least 50% of the room below (see Diagram1);

b. the distance between the foot of the access stair to the gallery and the door to the room containing the gallery should not exceed 3m;

c. the distance from the head of the access stair to any point on the gallery should not exceed 7.5m; and

d. any cooking facilities within a room containing a gallery should either:

   i. be enclosed with fire-resisting construction; or

   ii. be remote from the stair to the gallery and positioned such that they do not prejudice the escape from the gallery.

Diagram 1 Gallery floors with no alternative exit

See para 2.18

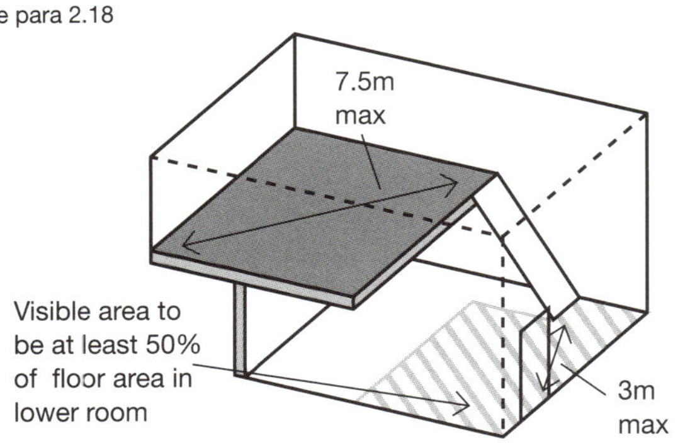

Notes:

1 This diagram does not apply where the gallery is:
   i. provided with an alternative escape route; or
   ii. provided with an emergency egress window (where the gallery floor is not more than 4.5m above ground level).

2 Any cooking facilities within a room containing a gallery should either:
   i. be enclosed with fire-resisting construction; or
   ii. be remote from the stair to the gallery and positioned such that they do not prejudice the escape from the gallery.

Emergency egress windows and external doors

2.9 Any window provided for emergency egress purposes should comply with the following conditions:

a. the window should have an unobstructed openable area that is at least 0.33m^2 and at least 450mm high and 450mm wide (the route through the window may be at an angle rather than straight through). The bottom of the openable area should be not more than 1100mm above the floor; and

b. the window should enable the person escaping to reach a place free from danger from fire.

Note 1: Approved Document K *Protection from falling, collision and impact* specifies a minimum guarding height of 800mm, except in the case of a window in a roof where the bottom of the opening may be 600mm above the floor.

Note 2: Locks (with or without removable keys) and stays may be fitted to egress windows, subject to the stay being fitted with a release catch, which may be child resistant.

Note 3: Windows should be designed such that they will remain in the open position without needing to be held by a person making their escape.

Provisions for escape from flats where the floor is not more than 4.5m above ground level

2.10 The internal arrangement of flats (single or multi-storey) where no floor is more than 4.5m in height may either meet the provisions in paragraphs 2.11 to 2.12 or 2.13 to 2.18.

Note: Where a flat is accesed via the common parts of a block of flats it may be necessary to provide a protected entrance hall to meet the provisions of Paragraph 2.21 and Diagram 9.

Provisions for escape from the ground storey

2.11 Except for kitchens, all habitable rooms in the ground storey should either:

a. open directly onto a hall leading to the entrance or other suitable exit; or

b. be provided with a window (or door) which complies with paragraph 2.9.

Provisions for escape from upper floors not more than 4.5m above ground level

2.12 Except for kitchens, all habitable rooms in the upper storey(s) should be provided with:

a. a window (or external door) which complies with paragraph 2.9; or

b. in the case of a multi-storey flat, direct access to its own internal protected stairway leading to a final exit.

Note: A single window can be accepted to serve two rooms provided both rooms have their own access to the stairs. A communicating door between the rooms must be provided so that it is possible to gain access to the window without passing through the stair enclosure.

Provisions for flats with a floor more than 4.5m above ground level

Internal planning of flats

2.13 Three acceptable approaches (all of which should observe the restrictions concerning inner rooms given in paragraph 2.5) when planning a flat which has a floor at more than 4.5m above ground level are:

a. to provide a protected entrance hall which serves all habitable rooms, planned so that the travel distance from the entrance door to the door to any habitable room is 9m or less (see Diagram 2); or

b. to plan the flat so that the travel distance from the entrance door to any point in any of the habitable rooms does not exceed 9m and the cooking facilities are remote from the entrance door and do not prejudice the escape route from any point in the flat (see Diagram 3); or

c. to provide an alternative exit from the flat, complying with paragraph 2.14.

Diagram 2 **Flat where all habitable rooms have direct access to an entrance hall**

See para 2.13(a)

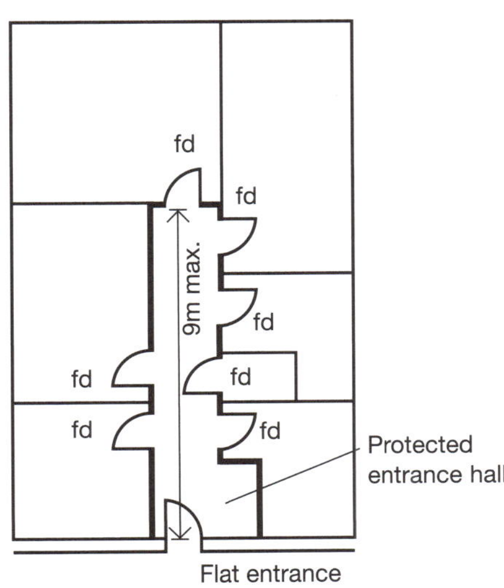

Note: Bathrooms need not have fire doors providing the bathroom is separated by fire-resisting construction from the adjacent rooms.

Key

fd Fire door

▬ 30 minute fire-resisting construction around entrance hall

Diagram 3 **Flat with restricted travel distance from furthest point to entrance**

See para 2.13(b)

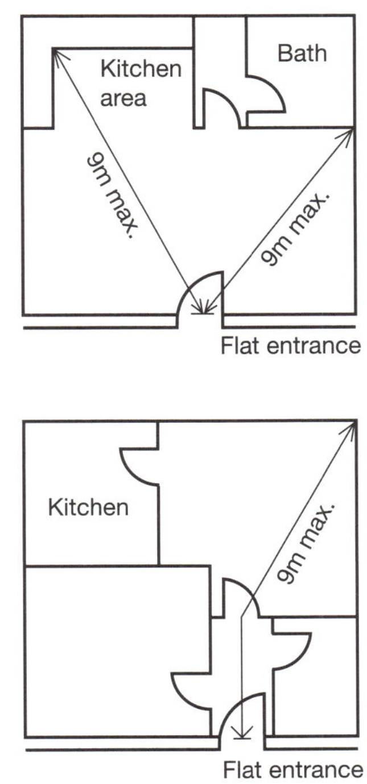

2.14 Where any flat has an alternative exit and the habitable rooms do not have direct access to the entrance hall (see Diagram 4):

a. the bedrooms should be separated from the living accommodation by fire-resisting construction and fire door(s); and

b. the alternative exit should be located in the part of the flat containing the bedroom(s).

Internal planning of flats with more than one storey

2.15 A multi-storey flat with an independent external entrance at ground level is similar to a dwellinghouse and means of escape should be planned on the basis of paragraphs 2.11 or 2.12 depending on the height of the top storey above ground level.

2.16 Four acceptable approaches to planning a multi-storey flat, which does not have its own external entrance at ground level but has a floor at more than 4.5m above ground level, are:

a. to provide an alternative exit from each habitable room which is not on the entrance floor of the flat, (see Diagram 5); or

b. to provide one alternative exit from each floor (other than the entrance floor), with a protected landing entered directly from all the habitable rooms on that floor, (see Diagram 6); or

c. where the vertical distance between the floor of the entrance storey and the floors above and below it does not exceed 7.5m, to provide a protected stairway plus additional smoke alarms in all habitable rooms and a heat alarm in any kitchen; or

d. to provide a protected stairway plus a sprinkler system in accordance paragraph 0.16 (smoke alarms should also be provided in accordance with paragraph 1.9).

Diagram 4 Flat with an alternative exit, but where all habitable rooms have no direct access to an entrance hall

See para 2.14

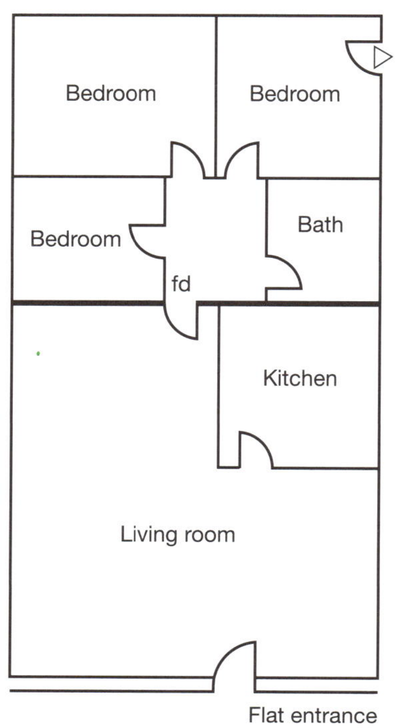

Note: The bedrooms are not classified as inner rooms because escape is possible in two directions.

Key
fd Fire door
▬ 30 minute fire-resisting construction between living and bedroom accommodation
△ Alternative exit

Diagram 5 Multi-storey flat with alternative exits from each habitable room, except at entrance level

See para 2.16(a)

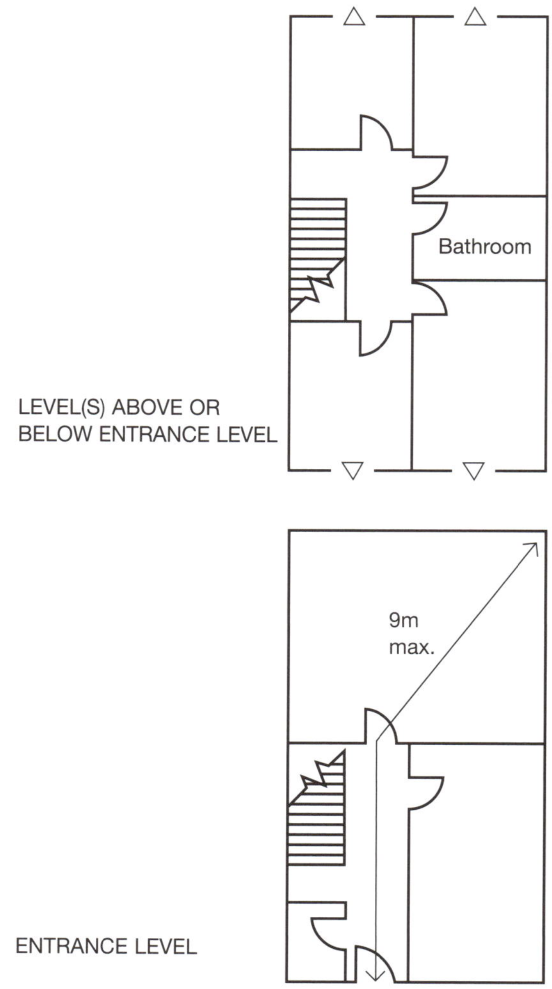

Note: This only applies where at least one storey is more than 4.5m above ground level.

Key
△ Alternative exit

Alternative exits

2.17 To be effective, an alternative exit from a flat should satisfy the following conditions:

a. be remote from the main entrance door to the flat; and

b. lead to a final exit or common stair by way of:

   i. a door onto an access corridor, access lobby or common balcony; or

   ii. an internal private stair leading to an access corridor, access lobby or common balcony at another level; or

   iii. a door into a common stair; or

   iv. a door onto an external stair; or

   v. a door onto an escape route over a flat roof.

Note: Any such access to a final exit or common stair should meet the appropriate provisions dealing with means of escape in the common parts of the building (see paragraph 2.19).

Diagram 6 Multi-storey flat with protected entrance hall and landing

See para 2.16(b)

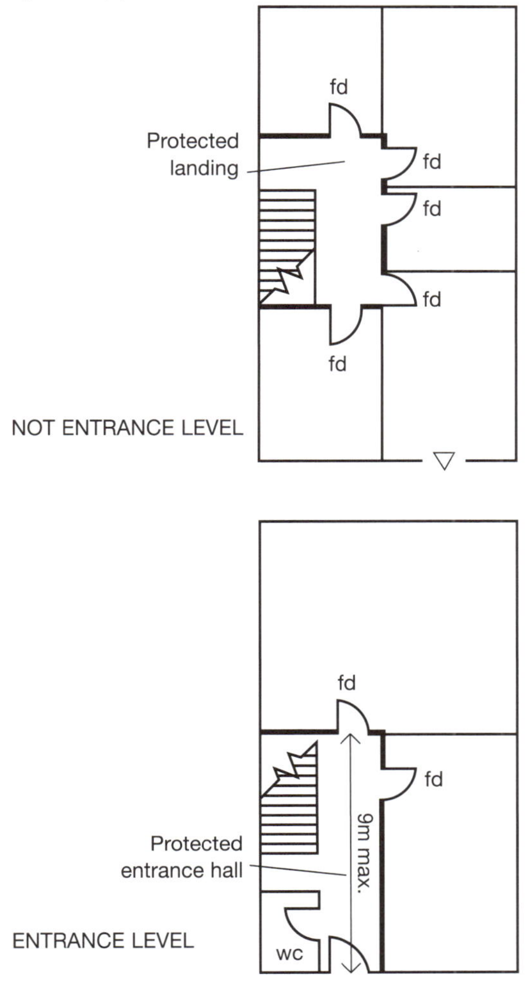

Note: This only applies where at least one storey is more than 4.5m above ground level.

Key

fd Fire door

▬ 30 minute fire-resisting stair enclosure

△ Alternative exit

Air circulation systems in flats with a protected stairway or entrance hall

2.18 Where ventilation ducts pass through compartment walls, then the guidance given paragraphs 5.46 to 5.53, 8.40 and 10.9 to 10.15 should be followed. Where an air circulation system circulates air only within an individual flat with an internal protected stairway or entrance hall the following precautions are needed to avoid the possibility of the system allowing smoke or fire to spread into the protected space.

a. Transfer grilles should not be fitted in any wall, door, floor or ceiling enclosing a protected stairway or entrance hall;

b. Any duct passing through through the enclosure to a protected stairway or entrance hall should of rigid steel construction and all joints between the ductwork and the enclosure should be fire-stopped,

c. Ventilation ducts supplying or extracting air directly to or from a protected stairway or entrance hall, should not also serve other areas;

d. Any system of mechanical ventilation which recirculates air and which serves both the stairway or entrance hall and other areas should be designed to shut down on the detection of smoke within the system; and

e. A room thermostat for a ducted warm air heating system should be mounted in the living room at a height between 1370mm and 1830mm and its maximum setting should not exceed 27°C.

Means of escape in the common parts of flats

2.19 The following paragraphs deal with means of escape from the entrance doors of flats to a final exit. They should be read in conjunction with the general provisions in Section 5.

Note: Paragraphs 2.20 to 2.51 are not applicable where the top floor is not more than 4.5m above ground level and the flats are designed in accordance with paragraphs 2.11 and 2.12. However, attention is drawn to the provisions in paragraph 0.29 regarding sheltered housing, Section 5 regarding general provisions, Section 8 (B3) regarding the provision of compartment walls and protected shafts and Section 16 (B5) regarding the provision of access for the Fire and Rescue Service.

Number of escape routes

2.20 Every flat should have access to alternative escape routes so that a person confronted by the effects of an outbreak of fire in another flat can turn away from it and make a safe escape.

However, a single escape route from the flat entrance door is acceptable if either:

a. the flat is situated in a storey served by a single common stair and:

   i. every flat is separated from the common stair by a protected lobby or common corridor (see Diagram 7); and

   ii. the travel distance limitations in Table 1 (see paragraph 2.23), on escape in one direction only, are observed; or

b. alternatively the flat is situated in a dead end part of a common corridor served by two (or more) common stairs and the travel distance complies with the limitations in Table 1 on escape in one direction only (see Diagram 8).

Diagram 7 **Flats served by one common stair**

See para 2.20(a) and 2.25

a. CORRIDOR ACCESS DWELLINGS

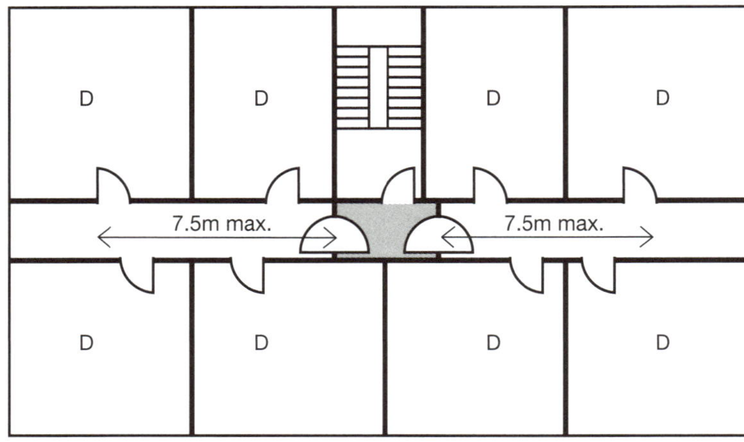

b. LOBBY ACCESS DWELLINGS

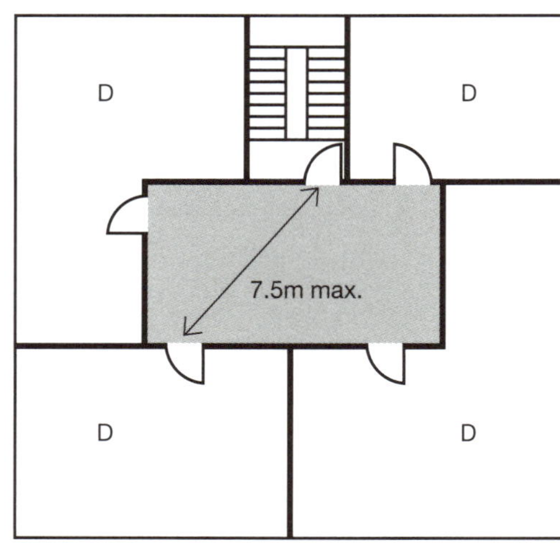

Note:

1. The arrangements shown also apply to the top storey.
2. See Diagram 9 for small single stair buildings.
3. All doors shown are fire doors.
4. Where travel distance is measured to a stair lobby, the lobby must not provide direct access to any storage room, flat or other space containing a potential fire hazard.

Key

D Dwelling

Shaded area indicates zone where ventilation should be provided in accordance with paragraph 2.26 (An external wall vent or smoke shaft located anywhere in the shaded area)

Diagram 8 **Flats served by more than one common stair**

See para 2.20(b) and 2.28

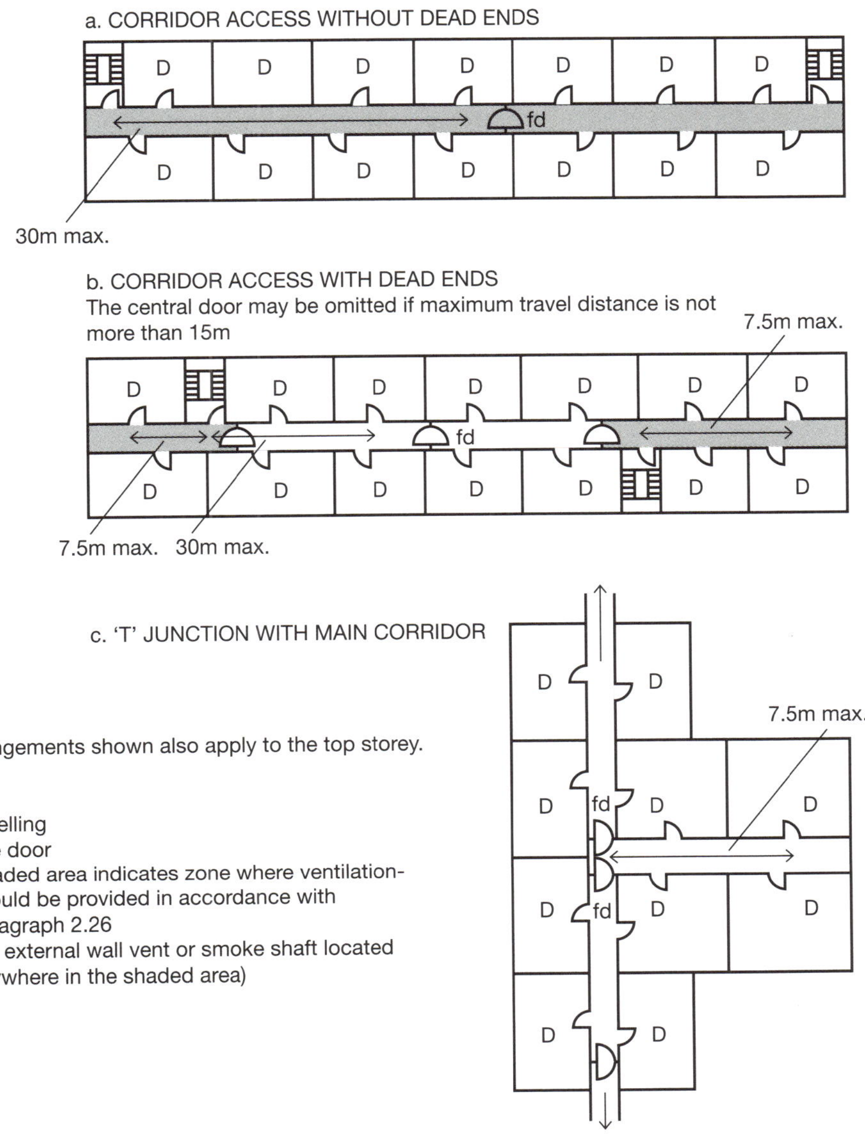

Note:
The arrangements shown also apply to the top storey.

Key
D Dwelling
fd Fire door
Shaded area indicates zone where ventilation-should be provided in accordance with paragraph 2.26
(An external wall vent or smoke shaft located anywhere in the shaded area)

Small single-stair buildings

2.21 The provisions in paragraph 2.20 may be modified and a single stair, protected in accordance with Diagram 9, may be used provided that:

a. the top floor of the building is no more than 11m above ground level;

b. there are no more than 3 storeys above the ground level storey;

c. the stair does not connect to a covered car park;

d. the stair does not serve ancillary accommodation unless the ancillary accommodation is separated from the stair by a protected lobby, or protected corridor, which has not less than $0.4m^2$ permanent ventilation or is protected from the ingress of smoke by a mechanical smoke control system; and

e. a high level openable vent, for fire and rescue service, use is provided at each floor level with a minimum free area of $1m^2$. Alternatively, a single openable vent may be provided at the head of the stair which can be remotely operated from fire and rescue service access level.

Flats with balcony or deck access

2.22 The provisions of paragraph 2.20 may also be modified in the case of flats with balcony or deck approach. Guidance on these forms of development is set out in clause 13 of BS 5588-1:1990.

Diagram 9 **Common escape route in small single stair building**

See para 2.21

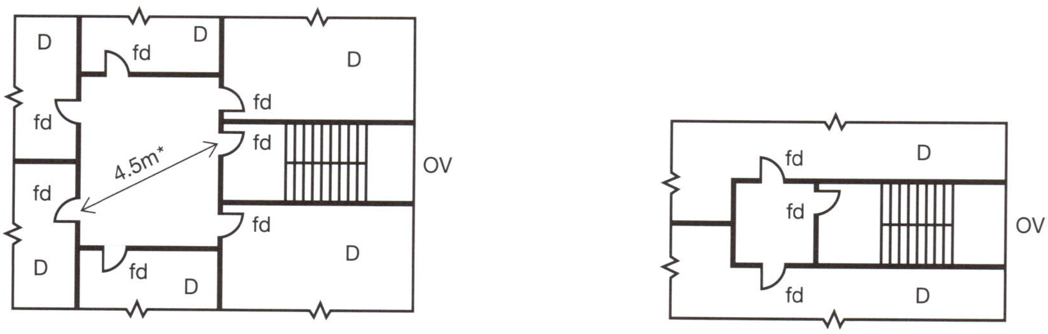

a. SMALL SINGLE STAIR BUILDING
*If smoke control is provided in the lobby, the travel distance can be increased to 7.5m maximum (see Diagram 7, example b).

b. SMALL SINGLE STAIR BUILDING WITH NO MORE THAN 2 DWELLINGS PER STOREY
The door between stair and lobby should be free from security fastenings.

If the dwellings have protected entrance halls, the lobby between the common stair and dwelling entrance is not essential.

Notes:

1. The arrangements shown also apply to the top storey.
2. If the travel distance across the lobby in Diagram 9a exceeds 4.5m, Diagram 7 applies.
3. Where, in Diagram 9b, the lobby between the common stair and the dwelling is omitted in small single-stair buildings, an automatic opening vent with a geometric free area of at least $1.0m^2$ is required at the top of the stair, to be operated on detection of smoke at any storey in the stair.

Key

▬ Fire-resisting construction
OV Openable vent at high level for fire service use ($1.0m^2$ minimum free area) see paragraph 2.21e
D Dwelling
fd Fire door

Table 1 **Limitations on distance of travel in common areas of blocks of flats** (see paragraph 2.23)

Maximum distance of travel (m) from flat entrance door to common stair, or to stair lobby (4)	
Escape in one direction only	Escape in more than one direction
7.5m (1)(2)	30m (2)(3)
Maximum distance of travel to storey exit in car parks	
Escape in one direction only	Escape in more than one direction
25m	45m

Notes:

1 Reduced to 4.5m in the case shown in Diagram 9.

2 Where all flats on a storey have independent alternative means of escape, the maximum distance of travel does not apply. However, see paragraph 16.3 (B5) which specifies Fire and Rescue Service access requirements.

3 For sheltered housing, see paragraph 0.29.

4 Where travel distance is measured to a stair lobby, the lobby must not provide direct access to any storage room, flat or other space containing a potential fire hazard.

Planning of common escape routes

2.23 Escape routes in the common areas should comply with the limitations on travel distance in Table 1. However, there may be circumstances where some increase on these maximum figures will be reasonable.

Escape routes should be planned so that people do not have to pass through one stairway enclosure to reach another. However, it is acceptable to pass through a protected lobby of one stairway in order to reach another.

Protection of common escape routes

2.24 To reduce the risk of a fire in a flat affecting the means of escape from other flats and common parts of the building, the common corridors should be protected corridors.

The wall between each flat and the corridor should be a compartment wall (see Section 8).

Smoke control of common escape routes

2.25 Despite the provisions described in this Approved Document, it is probable that some smoke will get into a common corridor or lobby from a fire in a flat, if only because the entrance door will be opened when the occupants escape.

There should therefore be some means of ventilating the common corridors/lobbies to control smoke and so protect the common stairs. This offers additional protection to that provided by the fire doors to the stair. (The ventilation also affords some protection to the corridors/lobbies.)

This can be achieved by either natural means in accordance with paragraph 2.26 or by means of mechanical ventilation as described in paragraph 2.27.

Smoke control of common escape routes by natural smoke ventilation

2.26 In buildings, other than small ones complying with Diagram 9, the corridor or lobby adjoining the stair should be provided with a vent. The vent from the corridor/lobby should be located as high as practicable and such that the top edge is at least as high as the top of the door to the stair.

There should also be a vent, with a free area of at least 1.0m², from the top storey of the stairway to the outside.

In single stair buildings the smoke vents on the fire floor and at the head of the stair should be actuated by means of smoke detectors in the common access space providing access to the flats. In buildings with more than one stair the smoke vents may be actuated manually (and accordingly smoke detection is not required for ventilation purposes). However, where manual actuation is used, the control system should be designed to ensure that the vent at the head of the stair will be opened either before, or at the same time, as the vent on the fire floor.

Vents should either:

a. be located on an external wall with minimum free area of 1.5m² (see Appendix C); or

b. discharge into a vertical smoke shaft (closed at the base) meeting the following criteria:

   i. Minimum cross-sectional area 1.5m² (minimum dimension 0.85m in any direction), opening at roof level at least 0.5m above any surrounding structures within a horizontal distance of 2.0m. The shaft should extend at least 2.5m above the ceiling of the highest storey served by the shaft;

   ii. The minimum free area of the vent from the corridor/lobby into the shaft and at the opening at the head of the shaft and at all internal locations within the shaft (e.g. safety grilles) should be at least 1.0m² (see Appendix C);

   iii. The smoke shaft should be constructed from non-combustible material and all vents should have a fire/smoke resistance performance at least that of an $E30S_a$ fire door. The shaft should be vertical from base to head, with no more than 4m at an inclined angle (maximum 30°); and

   iv. On detection of smoke in the common corridor/lobby, the vent(s) on the fire floor, the vent at the top of the smoke shaft and to the stairway should all open simultaneously. The vents from the corridors/lobbies on all other storeys should remain closed.

Smoke control of common escape routes by mechanical ventilation

2.27 As an alternative to the natural ventilation provisions in paragraph 2.26, mechanical ventilation to the stair and/or corridor/lobby may be provided to protect the stair(s) from smoke. Guidance on the design of smoke control systems using pressure differentials is available in BS EN 12101-6:2005.

Sub-division of common escape routes

2.28 A common corridor that connects two or more storey exits should be sub-divided by a self-closing fire door with, if necessary, any associated fire-resisting screen (see Diagram 8). The door(s) should be positioned so that smoke will not affect access to more than one stairway.

2.29 A dead-end portion of a common corridor should be separated from the rest of the corridor by a self-closing fire door with, if necessary, any associated fire-resisting screen (see Diagram 7a and Diagram 8b and 8c).

Ancillary accommodation, etc.

2.30 Stores and other ancillary accommodation should not be located within, or entered from, any protected lobby or protected corridor forming part of the only common escape route from a flat on the same storey as that ancillary accommodation.

Reference should be made to paragraphs 5.54 to 5.57 for special provisions for refuse chutes and storage areas.

Escape routes over flat roofs

2.31 If more than one escape route is available from a storey, or part of a building, one of those routes may be by way of a flat roof provided that it complies with the provisions in paragraph 5.35.

Note: Access to designs described in paragraph 2.48 may also be via a flat roof if the route over the roof complies with the provisions in paragraph 5.35.

Common stairs

Number of common stairs

2.32 As explained in paragraph 2.19 and paragraph 2.20 a single common stair can be acceptable in some cases, but otherwise there should be access to more than one common stair for escape purposes.

Width of common stairs

2.33 A stair of acceptable width for everyday use will be sufficient for escape purposes, but if it is also a firefighting stair, it should be at least 1100mm wide (see Appendix C for measurement of width).

Protection of common stairs

2.34 Common stairs need to have a satisfactory standard of fire protection if they are to fulfil their role as areas of relative safety during a fire evacuation. The provisions in paragraphs 2.35 to 2.46 should be followed.

2.35 Stairs provide a potential route for fire spread from floor to floor. In Section 7 under the requirement of B3 to inhibit internal fire spread, there is guidance on the enclosure of stairs to avoid this. A stair may also serve as a firefighting stair in accordance with the requirement B5, in which case account will have to be taken of the guidance in Section 17.

Enclosure of common stairs

2.36 Every common stair should be situated within a fire-resisting enclosure (i.e. it should be a protected stairway), to reduce the risk of smoke and heat making use of the stair hazardous.

2.37 The appropriate level of fire resistance is given in Appendix A, Tables A1 and A2.

Exits from protected stairways

2.38 Every protected stairway should discharge:

a. directly to a final exit; or

b. by way of a protected exit passageway to a final exit.

Note: Any such protected exit passageway should have the same standard of fire resistance and lobby protection as the stairway it serves.

Separation of adjoining protected stairways

2.39 Where two protected stairways (or exit passageways leading to different final exits) are adjacent, they should be separated by an imperforate enclosure.

Use of space within protected stairways

2.40 A protected stairway needs to be relatively free of potential sources of fire. Consequently, it should not be used for anything else, except a lift well or electricity meter(s). There are other provisions for lifts in paragraphs 5.39 to 5.45. In single stair buildings, meters located within the stairway should be enclosed within a secure cupboard which is separated from the escape route with fire-resisting construction.

Fire resistance and openings in external walls of protected stairways

2.41 The external enclosures to protected stairways should meet the provisions in paragraph 5.24.

Gas service and installation pipes in protected stairways

2.42 Gas service and installation pipes or associated meters should not be incorporated within a protected stairway unless the gas installation is in accordance with the requirements for installation and connection set out in the Pipelines Safety Regulations 1996, SI 1996 No 825 and the Gas Safety (Installation and Use) Regulations 1998 SI 1998 No 2451 (see also paragraph 8.40).

Basement stairs

2.43 Because of their situation, basement stairways are more likely to be filled with smoke and heat than stairs in ground and upper storeys.

Special measures are therefore needed in order to prevent a basement fire endangering upper storeys. These are set out in the following two paragraphs.

2.44 If an escape stair forms part of the only escape route from an upper storey of a building (or part of a building) which is not a small building (see paragraph 2.20), it should not be continued down to serve any basement storey. The basement should be served by a separate stair.

2.45 If there is more than one escape stair from an upper storey of a building (or part of a building), only one of the stairs serving the upper storeys of the building (or part) need be terminated at ground level. Other stairs may connect with the basement storey(s) if there is a protected lobby or a protected corridor between the stair(s) and accommodation at each basement level.

Stairs serving accommodation ancillary to flats

2.46 Except in small buildings described in paragraph 2.21, where a common stair forms part of the only escape route from a flat, it should not also serve any covered car park, boiler room, fuel storage space or other ancillary accommodation of similar fire risk.

2.47 Any common stair which does not form part of the only escape route from a flat may also serve ancillary accommodation if it is separated from the ancillary accommodation by a protected lobby or a protected corridor.

If the stair serves an enclosed (non open-sided) car park, or place of special fire hazard, the lobby or corridor should have not less than 0.4m^2 permanent ventilation or be protected from the ingress of smoke by a mechanical smoke control system.

External escape stairs

2.48 If the building (or part of the building) is served by a single access stair, that stair may be external if it:

a. serves a floor not more than 6m above the ground level; and

b. meets the provisions in paragraph 5.25.

2.49 Where more than one escape route is available from a storey (or part of a building), some of the escape routes from that storey or part of the building may be by way of an external escape stair, provided that there is at least one internal escape stair from every part of each storey (excluding plant areas) and the external stair(s):

a. serves a floor not more than 6m above either the ground level or a roof or podium which is itself served by an independent protected stairway; and

b. meets the provisions in paragraph 5.25.

Flats in mixed use buildings

2.50 In buildings with not more than three storeys above the ground storey, stairs may serve both flats and other occupancies, provided that the stairs are separated from each occupancy by protected lobbies at all levels.

2.51 In buildings with more than three storeys above the ground storey, stairs may serve both flats and other occupancies provided that:

a. the flat is ancillary to the main use of the building and is provided with an independent alternative escape route;

b. the stair is separated from any other occupancies on the lower storeys by protected lobbies (at those storey levels);

Note: The stair enclosure should have at least the same standard of fire resistance as stipulated in Table A2 for the elements of structure of the building (and take account of any additional provisions in Section 17 if it is a firefighting stair).

c. any automatic fire detection and alarm system with which the main part of the building is fitted also covers the flat;

d. any security measures should not prevent escape at all material times.

Note: Additional measures, including increased periods of fire resistance may be required between the flat and any storage area where fuels such as petrol and LPG are present.

Live/work units

2.52 Where a flat is intended to serve as a workplace for its occupants and for persons who do not live on the premises, the following additional fire precautions will be necessary:

a. The maximum travel distance to the flat entrance door or an alternative means of escape (not a window) from any part of the working area should not exceed 18m; and

b. Any windowless accommodation should have escape lighting which illuminates the route if the main supply fails. Standards for the installation of a system of escape lighting are given in BS 5266-1:2005.

Note: Where the unit is so large that the travel distance in a. cannot be met then the assumptions set out in paragraph 2.3 may no longer be valid. In such circumstances the design of the building should be considered on a case by case basis.

Section 3: Design for horizontal escape – buildings other than flats

Introduction

3.1 The general principle to be followed when designing facilities for means of escape is that any person confronted by an outbreak of fire within a building can turn away from it and make a safe escape. This Section deals with the provision of means of escape from any point to the storey exit of the floor in question, for all types of building. It should be read in conjunction with the guidance on the vertical part of the escape route in Section 4 and the general provisions in Section 5.

It should be noted that guidance in this Section is directed mainly at smaller, simpler types of buildings. Detailed guidance on the needs of larger, more complex or specialised buildings, can be found elsewhere (see paragraphs 0.21 to 0.35).

It should also be noted that although most of the information contained in this Section is related to general issues of design, special provisions apply to the layouts of certain institutional buildings (see paragraphs 3.38 onwards).

In the case of small shop, office, industrial, storage and other similar premises (ones with no storey larger than 280m^2 and having no more than 2 storeys plus a basement storey), the guidance in paragraph 3.32 to 3.37 may be followed instead of the other provisions in this Section.

Escape route design

Number of escape routes and exits

3.2 The number of escape routes and exits to be provided depends on the number of occupants in the room, tier or storey in question and the limits on travel distance to the nearest exit given in Table 2.

Note: It is only the distance to the nearest exit that should be so limited. Any other exits may be further away than the distances in Table 2.

3.3 In multi-storey buildings (see Section 4) more than one stair may be needed for escape, in which case every part of each storey will need to have access to more than one stair. This does not prevent areas from being in a dead-end condition provided that the alternative stair is accessible in case the first one is not usable.

3.4 In mixed-use buildings, separate means of escape should be provided from any storeys (or parts of storeys) used for Residential or Assembly and Recreation purposes.

Single escape routes and exits

3.5 In order to avoid occupants being trapped by fire or smoke, there should be alternative escape routes from all parts of the building.

However a single route is acceptable for:

a. parts of a floor from which a storey exit can be reached within the travel distance limit for travel in one direction set in Table 2 (see also paragraph 3.7). This is provided that, in the case of places of assembly and bars, no one room in this situation has an occupant capacity of more than 60 people or 30 people if the building is in Institutional use (Purpose Group 2a). The calculation of occupant capacity is described in Appendix C; or

b. a storey with an occupant capacity of not more than 60 people, where the limits on travel in one direction only are satisfied (see Table 2).

3.6 In many cases there will not be an alternative at the beginning of the route. For example, there may be only one exit from a room to a corridor, from which point escape is possible in two directions. This is acceptable provided that the overall distance to the nearest storey exit is within the limits for routes where there is an alternative and the 'one direction only' section of the route does not exceed the limit for travel where there is no alternative, see Table 2. Diagram 10 shows an example of a dead-end condition in an open storey layout.

Access control measures

3.7 Measures incorporated into the design of a building to restrict access to the building or parts of it should not adversely affect fire safety provisions.

Whilst it may be reasonable to secure some escape routes outside normal business hours, the measures left in place should be sufficient to allow safe evacuation of any persons left inside the building (see paragraph 5.11).

Table 2 **Limitations on travel distance**

Purpose group	Use of the premises or part of the premises		Maximum travel distance [1] where travel is possible in: One direction only (m)	More than one direction (m)
2(a)	Institutional		9	18
2(b)	Other residential:			
	a. in bedrooms [2]		9	18
	b. in bedroom corridors		9	35
	c. elsewhere		18	35
3	Office		18	45
4	Shop and commercial [3]		18 [4]	45
5	Assembly and recreation:			
	a. buildings primarily for disabled people		9	18
	b. areas with seating in rows		15	32
	c. elsewhere		18	45
6	Industrial [5]			
		Normal Hazard	25	45
		Higher Hazard	12	25
7	Storage and other non-residential [5]			
		Normal Hazard	25	45
		Higher Hazard	12	25
2–7	Place of special fire hazard [6]		9 [7]	18 [7]
2–7	Plant room or rooftop plant:			
	a. distance within the room		9	35
	b. escape route not in open air (overall travel distance)		18	45
	c. escape route in open air (overall travel distance)		60	100

Notes:

1. The dimensions in the Table are travel distances. If the internal layout of partitions, fittings, etc is not known when plans are deposited, direct distances may be used for assessment. The direct distance is taken as 2/3rds of the travel distance.
2. Maximum part of travel distance within the room. (This limit applies within the bedroom (and any associated dressing room, bathroom or sitting room, etc) and is measured to the door to the protected corridor serving the room or suite. Sub-item (b) applies from that point along the bedroom corridor to a storey exit.)
3. Maximum travel distances within shopping malls are given in BS 5588:Part 10. Guidance on associated smoke control measures is given in a BRE report *Design methodologies for smoke and heat exhaust ventilation* (BR 368).
4. BS 5588:Part 10 applies more restrictive provisions to units with only one exit in covered shopping complexes.
5. In industrial and storage buildings the appropriate travel distance depends on the level of fire hazard associated with the processes and materials being used. Higher hazard includes manufacturing, processing or storage of significant amounts of hazardous goods or materials, including: any compressed, liquefied or dissolved gas, any substance which becomes dangerous by interaction with either air or water, any liquid substance with a flash point below 65°C including whisky or other spirituous liquor, any corrosive substance, any oxidising agent, any substance liable to spontaneous combustion, any substance that changes or decomposes readily giving out heat when doing so, any combustible solid substance with a flash point less than 120° Celsius, any substance likely to spread fire by flowing from one part of a building to another.
6. Places of special fire hazard are listed in the definitions in Appendix E.
7. Maximum part of travel distance within the room/area.Travel distance outside the room/area to comply with the limits for the purpose group of the building or part.

Diagram 10 Travel distance in dead-end condition

See para 3.7

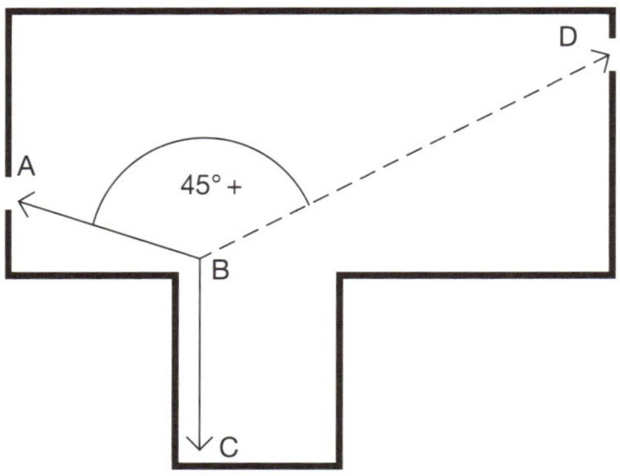

Angle ABD should be at least 45°. CBA or CBD (whichever is less) should be no more than the maximum distance of travel given for alternative routes and CB should be no more than the maximum distance for travel where there are no alternative routes.

Number of occupants and exits

3.8 The figure used for the number of occupants will normally be that specified as the basis for the design. When the number of occupants likely to use a room, tier or storey is not known, the capacity should be calculated on the basis of the appropriate floor space factors. Guidance for this is set out in Appendix C.

Table 3 gives the minimum number of escape routes and exits from a room or storey according to the number of occupants. (This number is likely to be increased by the need to observe travel distances and by other practical considerations.)

The width of escape routes and exits is the subject of paragraph 3.18.

Table 3 Minimum number of escape routes and exits from a room, tier or storey

Maximum number of persons	Minimum number of escape routes/exits
60	1
600	2
More than 600	3

Alternative escape routes

3.9 A choice of escape routes is of little value if they are all likely to be disabled simultaneously. Alternative escape routes should therefore satisfy the following criteria:

a. they are in directions 45° or more apart (see Diagram 11); or

b. they are in directions less than 45° apart, but are separated from each other by fire-resisting construction.

Diagram 11 Alternative escape routes

See para 3.9

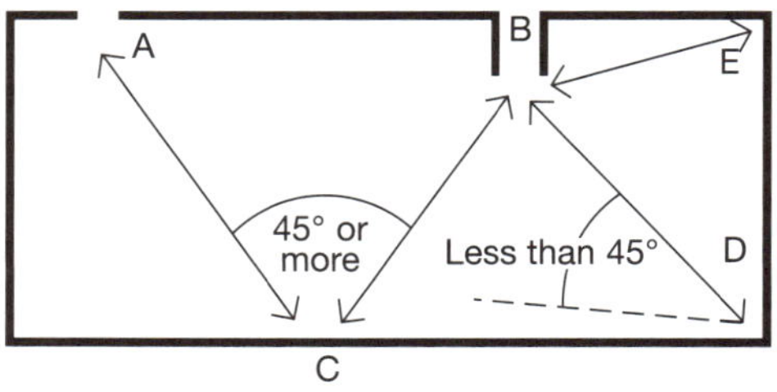

Alternative routes are available from C because angle ACB is 45° or more and therefore CA or CB (whichever is the less) should be no more than the maximum distance for travel given for alternative routes.

Alternative routes are not available from D because angle ADB is less than 45° (therefore see Diagram 10). There is also no alternative route from E.

Inner rooms

3.10 A room from which the only escape route is through another room is called an inner room. It is at risk if a fire starts in the other room, called the access room (see Diagram 12).

Such an arrangement is only acceptable if the following conditions are satisfied:

a. the occupant capacity of the inner room should not exceed 60 (30 in the case of a building in purpose group 2a (Institutional));

b. the inner room should not be a bedroom;

c. the inner room should be entered directly off the access room (but not via a corridor);

d. the escape route from the inner room should not pass through more than one access room;

e. the travel distance from any point in the inner room to the exit(s) from the access room should not exceed the appropriate limit given in Table 2;

f. the access room should not be a place of special fire hazard and should be in the control of the same occupier; and

g. one of the following arrangements should be made:

   i. the enclosures (walls or partitions) of the inner room should be stopped at least 500mm below the ceiling; or

   ii. a suitably sited vision panel not less than $0.1m^2$ should be located in the door or walls of the inner room, to enable occupants of the inner room to see if a fire has started in the outer room; or

   iii. the access room should be fitted with a suitable automatic fire detection and alarm system to warn the occupants of the inner room of the outbreak of a fire in the access room.

Diagram 12 **Inner room and access room**

See para 3.10

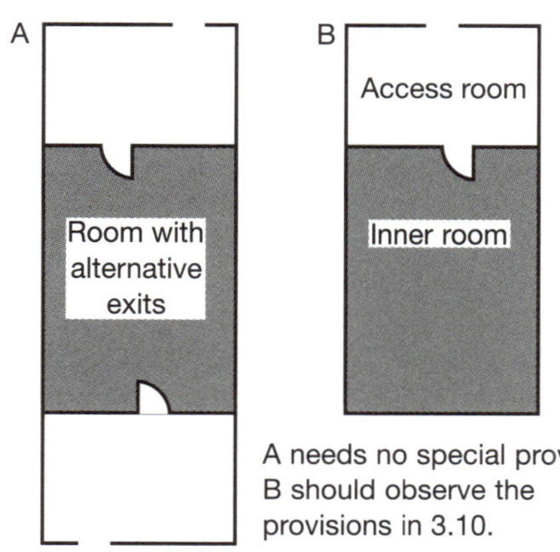

A needs no special provision.
B should observe the provisions in 3.10.

Planning of exits in a central core

3.11 Buildings with more than one exit in a central core should be planned so that storey exits are remote from one another and so that no two exits are approached from the same lift hall, common lobby or undivided corridor, or linked by any of these (see Diagram 13).

Open spatial planning

3.12 Escape routes should not be prejudiced by openings between floors, such as an escalator. (see Diagram 14).

An escape route should not be within 4.5m of the openings unless:

a. the direction of travel is away from the opening; or

b. there is an alternative escape route which does not pass within 4.5m of the open connection.

Diagram 13 **Exits in a central core**

See para 3.11

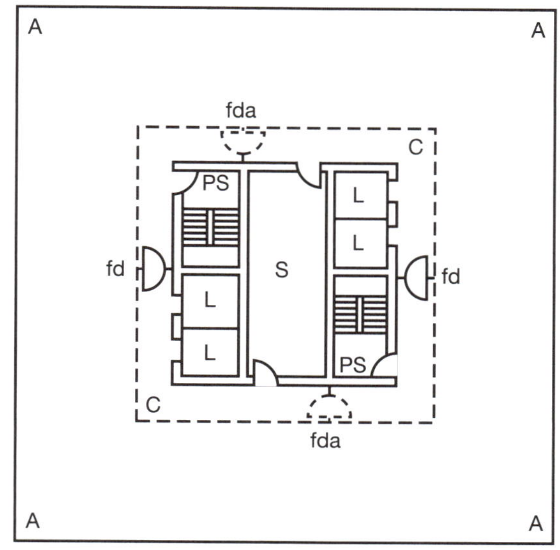

Note: The doors at both ends of the area marked 'S' should be self-closing fire doors unless the area is sub-divided such that any fire in that area will not be able to prejudice both sections of corridor at the same time. If that area is a lift lobby, doors should be provided as shown in Figure 8 in BS 5588: Part 11: 1997.

Key

L Lift
S Services, toilets, etc.
fd Self-closing FD20S fire doors
fda Possible alternative position for fire door
C Corridor off which accommodation opens
PS Protected stairway
A Accommodation (e.g. office space)

Diagram 14 **Open connections**

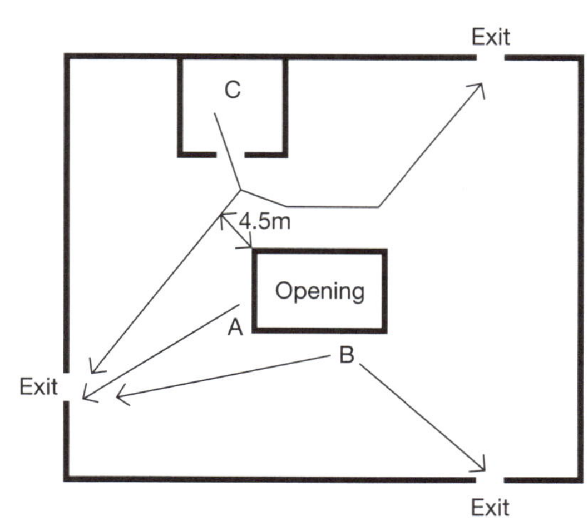

From A and B at least one direction of travel is away from the opening. From C where the initial direction of travel is towards the opening, one of the escape routes is not less than 4.5m from the opening.

Access to storey exits

3.13 Any storey which has more than one escape stair should be planned so that it is not necessary to pass through one stairway to reach another. However it would be acceptable to pass through one stairway's protected lobby to reach another stair.

Separation of circulation routes from stairways

3.14 Unless the doors to a protected stairway and any associated exit passageway are fitted with an automatic release mechanism (see Appendix B, paragraph 3b), the stairway and any associated exit passageway should not form part of the primary circulation route between different parts of the building at the same level. This is because the self-closing fire doors are more likely to be rendered ineffective as a result of their constant use, or because some occupants may regard them as an impediment. For example, the doors are likely to be wedged open or have their closers removed.

Storeys divided into different uses

3.15 Where a storey contains an area (which is ancillary to the main use of the building) for the consumption of food and/or drink, then:

a. not less than two escape routes should be provided from each such area (except inner rooms which meet the provisions in paragraph 3.10); and

b. the escape routes from each such area should lead directly to a storey exit without entering any kitchen or similar area of high fire hazard.

Storeys divided into different occupancies

3.16 Where any storey is divided into separate occupancies (i.e. where there are separate ownerships or tenancies of different organisations):

a. the means of escape from each occupancy should not pass through any other occupancy; and

b. if the means of escape include a common corridor or circulation space, either it should be a protected corridor, or a suitable automatic fire detection and alarm system should be installed throughout the storey.

Height of escape routes

3.17 All escape routes should have a clear headroom of not less than 2m except in doorways.

Width of escape routes and exits

3.18 The width of escape routes and exits depends on the number of persons needing to use them. They should not be less than the dimensions given in Table 4. (Attention is also drawn to the guidance in Approved Document M *Access to and Use of buildings*).

3.19 Where the maximum number of people likely to use the escape route and exit is not known, the appropriate capacity should be calculated on the basis of the occupant capacity. Guidance is set out in Appendix C.

3.20 Guidance on the spacing of fixed seating for auditoria is given in BS 5588-6:1991.

Table 4 Widths of escape routes and exits

Maximum number of persons	Minimum width mm (1) (2) (3)
60	750 (4)
110	850
220	1050
More than 220	5 per person (5)

Notes:

1. Refer to Appendix C on methods of measurement.
2. In order to follow the guidance in the Approved Document to Part M the widths given in the table may need to be increased.
3. Widths less than 1050mm should not be interpolated.
4. May be reduced to 530mm for gangways between fixed storage racking, other than in public areas of Purpose Group 4 (shop and commercial).
5. 5mm/person does not apply to an opening serving less than 220 persons.

Calculating exit capacity

3.21 If a storey or room has two or more storey exits it has to be assumed that a fire might prevent the occupants from using one of them. The remaining exit(s) need to be wide enough to allow all the occupants to leave quickly. Therefore when deciding on the total width of exits needed according to Table 4, the largest exit should be discounted. This may have implications for the width of stairs, because they should be at least as wide as any storey exit leading onto them. Although some stairs are not subject to discounting (see paragraphs 4.20 and 4.21), the storey exits onto them will be.

3.22 The total number of persons which two or more available exits (after discounting) can accommodate is found by adding the maximum number of persons that can be accommodated by each exit width. For example, 3 exits each 850mm wide will accommodate 3 x 110 = 330 persons (**not** the 510 persons accommodated by a single exit 2550mm wide).

3.23 Where a ground floor storey exit shares a final exit with a stair via a ground floor lobby, the width of the final exit should be sufficient to enable a maximum evacuation flow rate equal to or greater than that from the storey exit and stair combined (see Diagram 15).

Diagram 15 **Merging flows at final exit**

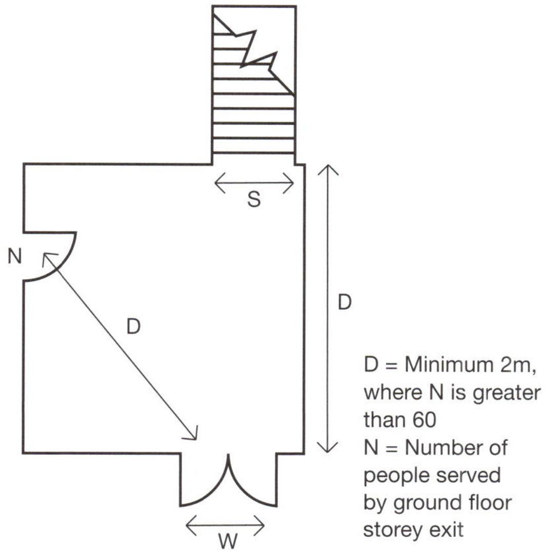

This can be calculated from the following formula:

W = ((N/2.5) + (60S))/80

Where:

W = width of final exit, in metres

N = number of people served by ground floor storey exit

S = stair width in metres

Note: Where the number of persons (N) entering the lobby from the ground floor is more than 60 then the distance from the foot of the stair, or the storey exit, to the final exit should be a minimum of two metres (see Diagram 15). Where this cannot be achived then the width of the final exit (W) should be no less than the width of the stair plus the width of the storey exit.

Worked example

A ground floor storey exit serving 250 persons shares a common final exit with a 1.2 m wide stair

Required final exit = ((250/2.5) + (1.2 x 60))/80 width (metres) = 2.150 metres

Protected corridors

3.24 A corridor which serves a part of the means of escape in any of the following circumstances should be a protected corridor:

a. every corridor serving bedrooms;

b. every dead-end corridor (excluding recesses and extensions not exceding 2m deep as shown in Figures 10 and 11 of BS 5588-11:1997*);* and

c. any corridor common to two or more different occupancies (see also paragraph 3.16).

Enclosure of corridors that are not protected corridors

3.25 Where a corridor that is used as a means of escape, but is not a protected corridor, is enclosed by partitions, those partitions provide some defence against the spread of smoke in the early stages of a fire, even though they may have no fire resistance rating. To maintain this defence the partitions should be carried up to the soffit of the structural floor above, or to a suspended ceiling and openings into rooms from the corridor should be fitted with doors, which need not be fire doors. Open planning, while offering no impediment to smoke spread, has the compensation that occupants can become aware of a fire quickly.

Sub-division of corridors

3.26 If a corridor provides access to alternative escape routes, there is a risk that smoke will spread along it and make both routes impassable before all occupants have escaped.

To avoid this, every corridor more than 12m long which connects two or more storey exits, should be sub-divided by self-closing fire doors (and any necessary associated screens). The fire door(s) and any associated screen(s) should be positioned approximately mid-way between the two storey exits to effectively safeguard the route from smoke (having regard to the layout of the corridor and to any adjacent fire risks).

In a building of Purpose Groups 2 to 7, where a cavity exists above the enclosures to any such corridor, because the enclosures are not carried to full storey height or (in the case of a top storey) to the underside of the roof covering, the potential for smoke to bypass the sub-division should be restricted by:

a. fitting cavity barriers on the line of the enclosure(s) to and across the corridor (see Diagram 16a); or

b. sub-dividing the storey using fire-resisting construction passing through the line of the sub-division of the corridor (see diagram 16b). Any void above this subdivision should be fitted with cavity barriers on the line of sub-division of the storey and the corridor; or

c. enclosing the cavity on the lower side by a fire-resisting ceiling which extends throughout the building, compartment or separated part.

Any door which could provide a path for smoke to bypass the sub-division should be made self closing (but need not necessarily be fire-resisting).

Diagram 16 **Subdivision of corridors**

a. SECTION TO SHOW USE OF CAVITY BARRIERS ABOVE THE CORRIDOR ENCLOSURE see paragraph 3.26 a.

b. PLAN SHOWING SUB-DIVISION OF THE STOREY BY FIRE-RESISTING CONSTRUCTION See paragraph 3.26 b.

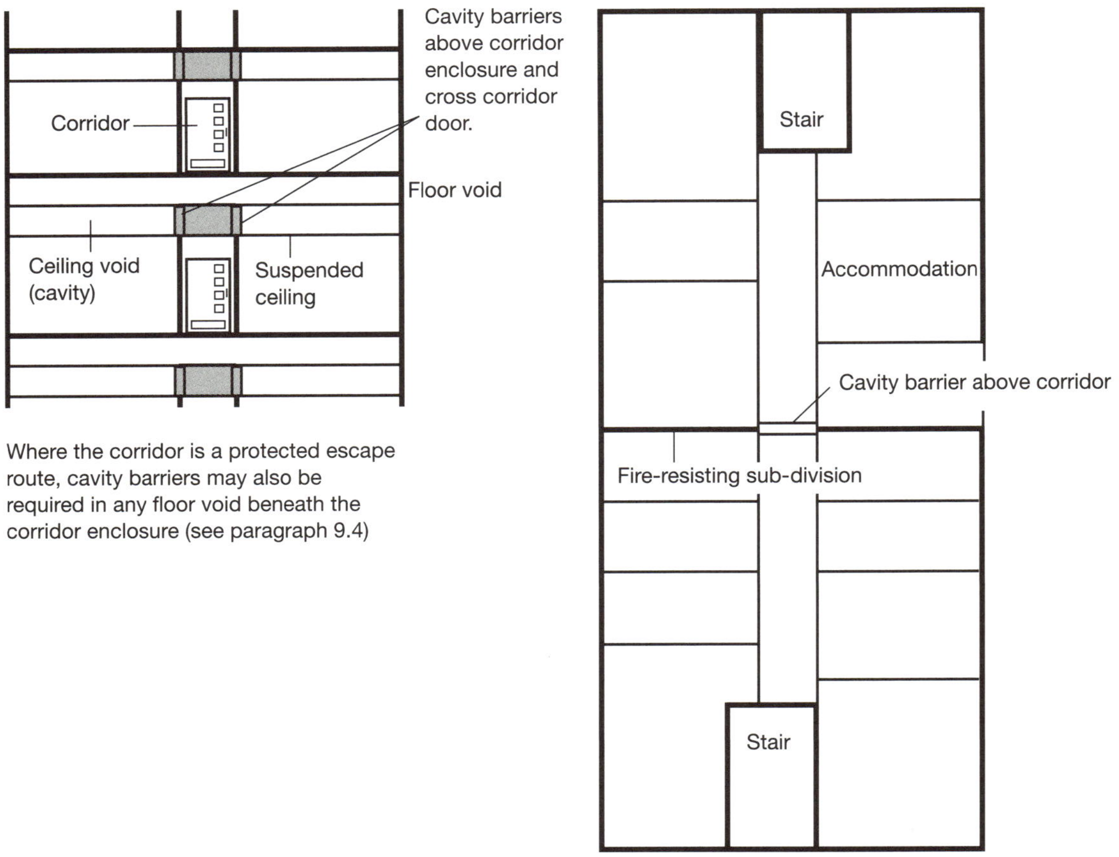

Where the corridor is a protected escape route, cavity barriers may also be required in any floor void beneath the corridor enclosure (see paragraph 9.4)

The sub-division should be carried full storey height and includes sub-division of the corridor. A cavity barrier may be used in any ceiling void over the sub-division.

3.27 If a dead-end portion of a corridor provides access to a point from which alternative escape routes are available, there is a risk that smoke from a fire could make both routes impassable before the occupants in the dead-end have escaped.

To avoid this, unless the escape stairway(s) and corridors are protected by a pressurization system complying with BS EN 12101-6:2005, every dead-end corridor exceeding 4.5m in length should be separated by self-closing fire doors (together with any necessary associated screens) from any part of the corridor which:

a. provides two directions of escape (see Diagram 17(a)); or

b. continues past one storey exit to another (see Diagram 17(b)).

Cavity barriers

3.28 Additional measures to safeguard escape routes from smoke are given in Section 9 (B3).

External escape routes

3.29 Guidance on the use of external escape stairs from buildings is given in paragraph 4.44.

3.30 Where an external escape route (other than a stair) is beside an external wall of the building, that part of the external wall within 1800mm of the escape route should be of fire-resisting construction, up to a height of 1100mm above the paving level of the route. For guidance on external escape stairs see paragraph 5.25.

Escape over flat roofs

3.31 If more than one escape route is available from a storey, or part of a building, one of those routes may be by way of a flat roof, provided that:

a. the route does not serve an Institutional building, or part of a building intended for use by members of the public; and

b. it meets the provisions in paragraph 5.35.

Diagram 17 **Dead-end corridors**

See para 3.26

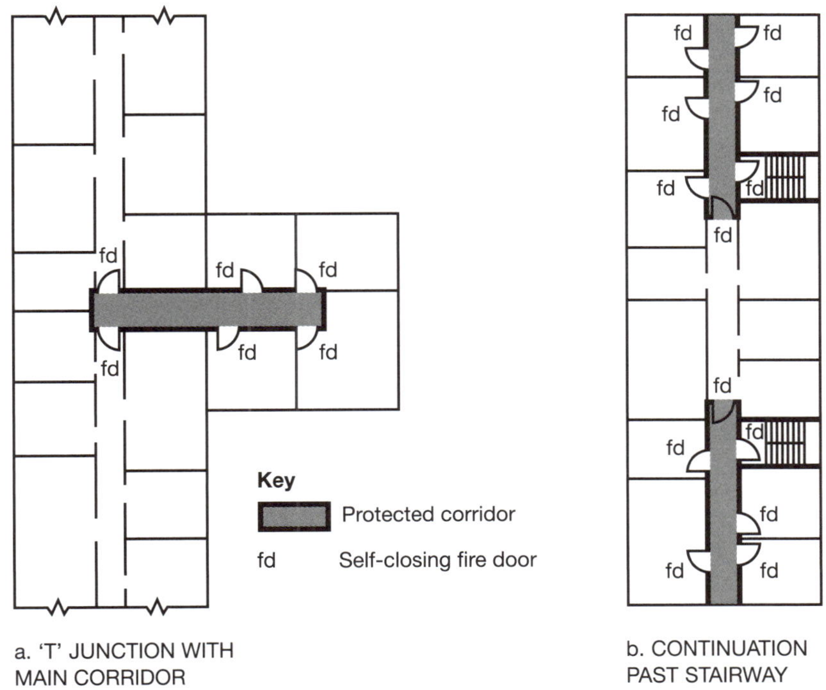

Small premises

3.32 In small premises, as described in paragraph 3.33, the number of persons is generally limited and the size of the premises, when undivided, will tend to enable clear vision of all parts. Therefore the occupants will be able to quickly reach an entrance/exit in an emergency. Thus a reduction in the number of exits and stairs as set out in paragraphs 3.33 to 3.37, 4.6 and 4.33 is acceptable. However, where the sale, storage or use of highly flammable materials is involved, it is necessary for persons to rapidly vacate the premises in the event of a fire. To facilitate this, the general guidance in paragraph 3.33 would not apply.

General

3.33 The following paragraphs apply in place of only those provisions relating to the number and positioning of exits and protected stairways and measurement of distances of travel.

Note 1: They do not apply to premises used principally for the storage and/or sale of highly flammable liquids or materials.

Note 2: In covered shopping complexes, the size of small units that may be served by a single exit is further restricted. This is dealt with in BS 5588-10:1991.

a. The premises should be in a single occupancy and should not comprise more than a basement, a ground floor and a first storey. No storey should have a floor area greater than 280m^2 (see Diagram 18);

b. Any kitchen or other open cooking arrangement should be sited at the extremity of any dead end remote from the exit(s); and

c. The planned seating accommodation or the assessed standing accommodation (see Table C1) for small premises comprising a bar or restaurant should not exceed 30 persons per storey. This figure may be increased to 100 persons for the ground storey if that storey has an independent final exit.

Diagram 18 **Maximum travel distances in a small two or three storey premises with a single protected stair to each storey**

See para 3.33

a. FIRST STOREY

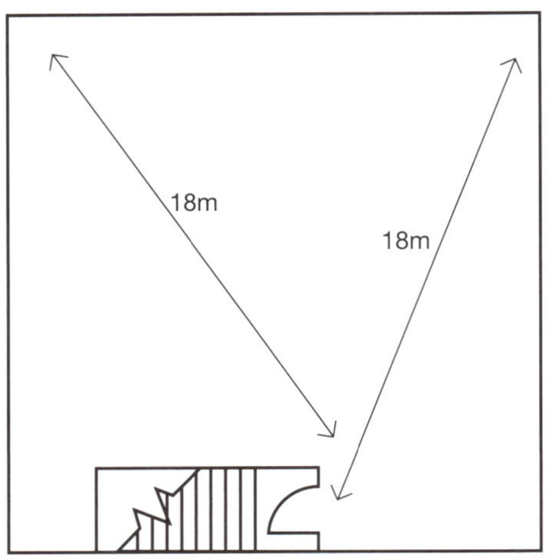

b. GROUND STOREY

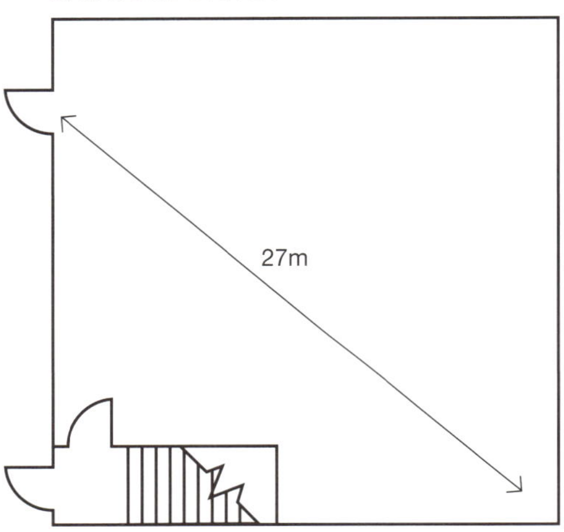

c. BASEMENT

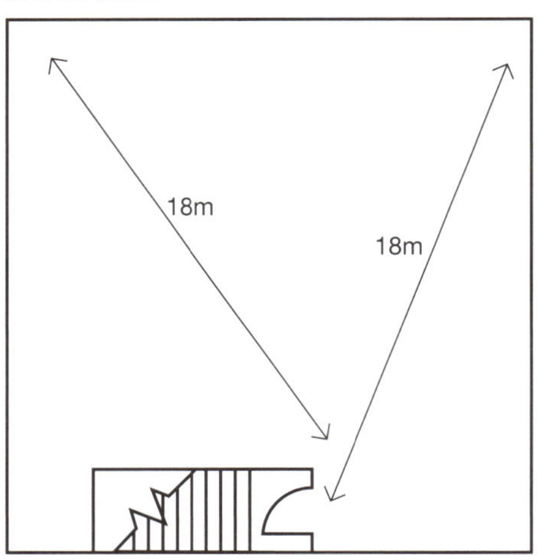

Note: Maximum floor area in any one storey 280m^2. Restricted accommodation if used as a restaurant or bar.

Construction

3.34 The floor areas should be generally undivided (except for kitchens, ancillary offices and stores) to ensure that exits are clearly visible from all parts of the floor areas.

3.35 Store rooms should be enclosed with fire-resisting construction.

3.36 Sufficient clear glazed areas should be provided in any partitioning separating a kitchen or ancillary office from the open floor area to enable any person within the kitchen or office to obtain early visual warning of an outbreak of fire. Alternatively, an automatic fire detection and alarm system may be provided in the outer room.

Note: The clear glazed area or vision panel may need to be provided for other reasons.

Distance of travel and number of escape routes

3.37 The escape routes from any storey should be of such a number and so situated that the distance of travel from any point to the nearest storey exit does not exceed the appropriate limits set out in Table 5.

Note: The distance of travel in small premises with an open stairway is measured to the foot of the stair in a basement or to the head of the stair in a first storey see paragraph 4.33.

The siting of two or more exits or stairs should be such that they afford effective alternative directions of travel from any relevant point in a storey.

Table 5 **Maximum distances of travel in small premises with a protected stair**

Storey	Maximum Travel Distance
Ground storey with a single exit	27
Basement or first storey with a single stair	18
Storey with more than one exit/stair	45

Note:

The dimensions in the Table are travel distances. If the internal layout of partitions, fittings, etc is not known when plans are deposited, direct distances may be used for assessment. The direct distance is taken as 2/3rds of the travel distance.

Residential care homes

General

3.38 Residential care homes are quite diverse and can be used by a variety of residents, often requiring different types of care to suit their specific needs. They can include homes for the elderly, children and people who are physically or mentally disabled. The choice of fire safety strategy is dependent upon the way a building is designed, furnished, staffed and managed and the level of dependency of the residents.

3.39 Generally, in care homes for the elderly it is reasonable to assume that at least a proportion of the residents will need some assistance to evacuate. As such these buildings should be designed for progressive horizontal evacuation (PHE) in accordance with paragraphs 3.41 to 3.52 below. For other types of care home a judgement should be made as to whether PHE or a simultaneous evacuation strategy is appropriate. Whatever approach is adopted in the design of a building this must be recorded and communicated to the building management to ensure that procedures are adopted that are compatible with the building design.

3.40 The guidance on PHE given in paragraphs 3.41 to 3.52 is for those care homes, to which the provisions of the *"Firecode"* documents are not applicable (see Para 0.23).

Planning for progressive horizontal evacuation

3.41 The concept of PHE requires those areas used for the care of residents to be subdivided into protected areas separated by compartment walls and compartment floors. This allows horizontal escape to be made by evacuating into adjoining protected areas. The objective is to provide a place of relative safety within a short distance, from which further evacuation can be made if necessary but under less pressure of time.

3.42 Each storey, used for the care of residents, should be divided into at least three protected areas by compartment walls and all floors should be compartment floors.

3.43 Every protected area should be provided with at least two exits to adjoining, but separate protected areas. Travel distances within a protected area to these exits should not exceed those given in Table 2. The maximum travel distance from any point should be not more than 64m to a storey exit or a final exit.

3.44 A fire in any one protected area should not prevent the occupants of any other area from reaching a final exit (see Diagram 19). Escape routes should not pass through ancillary accommodation such as that listed in paragraph 3.50.

3.45 The number of residents' beds in protected areas should be established based on assessment of the number of staff likely to be available and the level of assistance that residents may require. In no case should this exceed 10 beds in any one protected area.

3.46 Adjoining protected areas into which horizontal evacuation may take place should each have a floor area sufficient to accommodate not only their own occupants but also the occupants from the largest adjoining protected area.

Diagram 19 **Progressive horizontal evacuation in care homes**

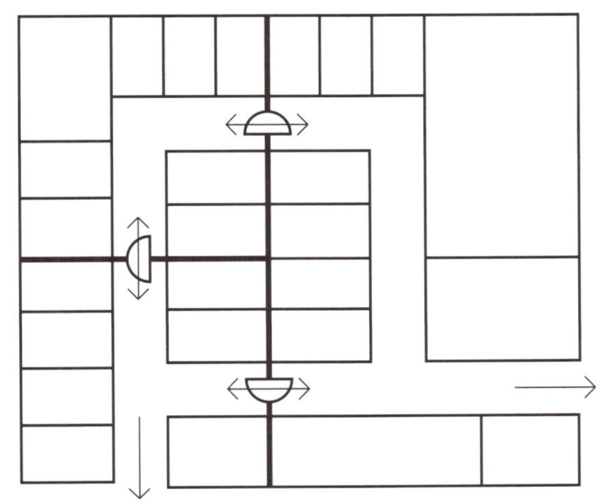

→ Escape route to adjacent compartment, storey exit or final exit.

▬ Compartment wall

Note: Bedrooms and and all ancillary accommodation should be enclosed in fire-resisting construction.

Fire detection

3.47 A fire detection and alarm system should be provided to an L1 standard in accordance with BS 5839-1:2002.

Bedrooms

3.48 Each bedroom should be enclosed in fire-resisting construction with fire resisting doors and every corridor serving bedrooms should be a protected corridor (see paragraph 3.24).

3.49 Bedrooms should not contain more than one bed (this includes a double bed).

Ancillary accommodation

3.50 Ancillary accomodation such as the following, should be enclosed by fire-resisting construction.

a. chemical stores;

b. cleaners' rooms;

c. clothes' storage;

d. day rooms;

e. smoking rooms;

f. disposal rooms;

g. plant rooms;

h. linen stores;

i. kitchens;

j. laundry rooms;

k. staff changing and locker rooms; and

l. store rooms.

Door-closing devices

3.51 The specification of door-closing devices for fire doors should take account of the needs of residents. In particular where self-closing doors could present an obstacle to the residents of the building then the following hardware in accordance with BS EN 1155:1997 would be appropriate;

Bedrooms – free-swing door closers.

Circulation spaces – hold-open devices.

Sprinkler systems

3.52 Where a sprinkler system is provided in accordance with paragraph 0.16. The following variations to the guidance given in paragraphs 3.41 to 3.51 are acceptable,

a. Fire doors to bedrooms need not be fitted with self closing devices.

b. Protected areas may contain more than 10 beds.

c. Bedrooms may contain more than one bed.

Note: Management procedures will need to take account of the larger number of residents that may need assistance and the need to manually close bedroom doors during sleeping hours.

Section 4: Design for vertical escape – buildings other than flats

Introduction

4.1 An important aspect of means of escape in multi-storey buildings is the availability of a sufficient number of adequately sized and protected escape stairs. This Section deals with escape stairs and includes measures necessary to protect them in all types of building.

The limitation of distances of horizontal travel for means of escape purposes means that most people should be able independently to reach the safety of a protected escape route or final exit. However, some people, for example those who use wheelchairs, may not be able to use stairways without assistance. For them evacuation involving the use of refuges on escape routes and either assistance down (or up) stairways, or the use of suitable lifts, will be necessary.

This Section should be read in conjunction with the general provisions in Section 5.

Number of escape stairs

4.2 The number of escape stairs needed in a building (or part of a building) will be determined by:

a. the constraints imposed in Section 3 on the design of horizontal escape routes;

b. whether independent stairs are required in mixed occupancy buildings (see paragraph 4.4);

c. whether a single stair is acceptable (see paragraphs 4.5 and 4.6); and

d. provision of adequate width for escape (see paragraph 4.15) while allowing for the possibility that a stair may have to be discounted because of fire or smoke (see paragraph 4.20).

4.3 In larger buildings, provisions for access for the Fire and Rescue Service may apply, in which case, some escape stairs may also need to serve as firefighting stairs. The number of escape stairs may therefore be affected by provisions made in Section 17, paragraphs 17.8 and 17.9.

Mixed use buildings

4.4 Where a building contains storeys (or parts of storeys) in different purpose groups, it is important to consider the effect of one risk on another. A fire in a shop, or unattended office, could have serious consequences on, for example, a residential or hotel use in the same building. It is therefore important to consider whether completely separate routes of escape should be provided from each different use within the building or whether other effective means to protect common escape routes can be provided.

Single escape stairs

4.5 Provided that independent escape routes are not necessary from areas in different purpose groups in accordance with paragraph 2.50 or 4.4, the situations where a building (or part of a building) may be served by a single escape stair are:

a. from a basement which is allowed to have a single escape route in accordance with paragraph 3.5b and Table 2;

b. from a building (other than small premises, see 4.5c) which has no storey with a floor level more than 11m above ground level and in which every storey is allowed to have a single escape route in accordance with paragraph 3.5b and Table 2;

c. in the case of small premises (see paragraph 3.32), in situations where the guidance in paragraph 4.6 is followed.

Single escape stairs in small premises

4.6 A single escape stair may be used from:

a. small premises as described in paragraph 3.33;

b. an office building comprising not more than five storeys above the ground storey, provided that:

   i. the travel distance from every point in each storey does not exceed that given in Table 2 for escape in one direction only; and

   ii. every storey at a height greater than 11m has an alternative means of escape;

c. a factory comprising not more than:

   i. two storeys above the ground storey (if the building, or part of the building, is of low risk); or

   ii. one storey above the ground storey (if the building, or part of the building, is of normal risk); provided that the travel distance from every point on each storey does not exceed that given in Table 2 for escape in one direction only; or

d. process plant buildings with an occupant capacity of not more than 10.

Provision of refuges

4.7 Refuges are relatively safe waiting areas for short periods. They are not areas where disabled people should be left alone indefinitely until rescued by the fire and rescue service, or until the fire is extinguished.

A refuge should be provided for each protected stairway affording egress from each storey, except storeys consisting exclusively of plant rooms.

Note: Whilst a refuge should be provided for each stairway, they need not necessarily be located within the stair enclosure but should enable direct access to the stair. The number of refuge spaces need not necessarily equal the sum of the number of wheelchair users who can be present in the building. Refuges form a part of the management plan and it may be that more than one disabled person will use a single refuge as they pass through as a part of the evacuation procedure.

4.8 The following are examples of satisfactory refuges:

a. an enclosure such as a compartment (see Diagram 20), protected lobby, protected corridor or protected stairway (see Diagram 21); and

b. an area in the open air such as a flat roof, balcony, podium or similar place which is sufficiently protected (or remote) from any fire risk and provided with its own means of escape.

4.9 Each refuge should provide an area accessible to a wheelchair of at least 900mm x 1400mm in which a wheelchair user can await assistance. Where a refuge is a protected stairway or protected lobby or protected corridor, the wheelchair space should not reduce the width of the escape route. Where the wheelchair space is within a protected stairway, access to the wheelchair space should not obstruct the flow of persons escaping.

Diagram 21 **Refuge formed in a protected stairway**

See para 4.8

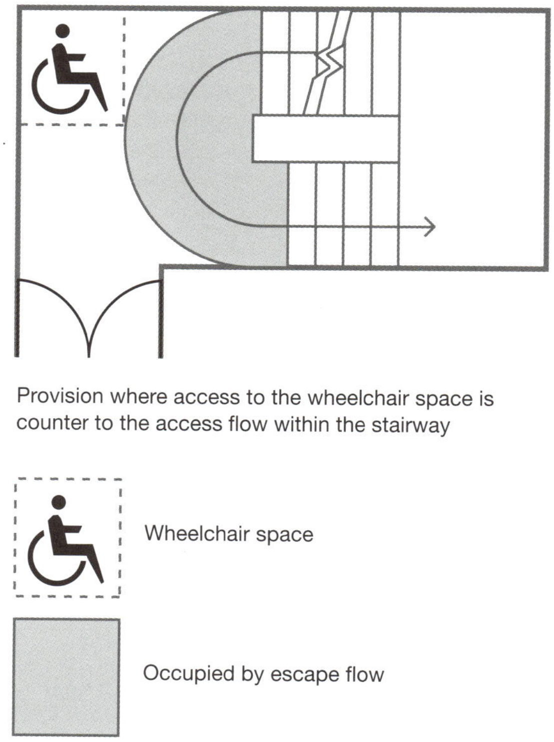

Diagram 20 **Refuge formed by compartmentation**

See para 4.8

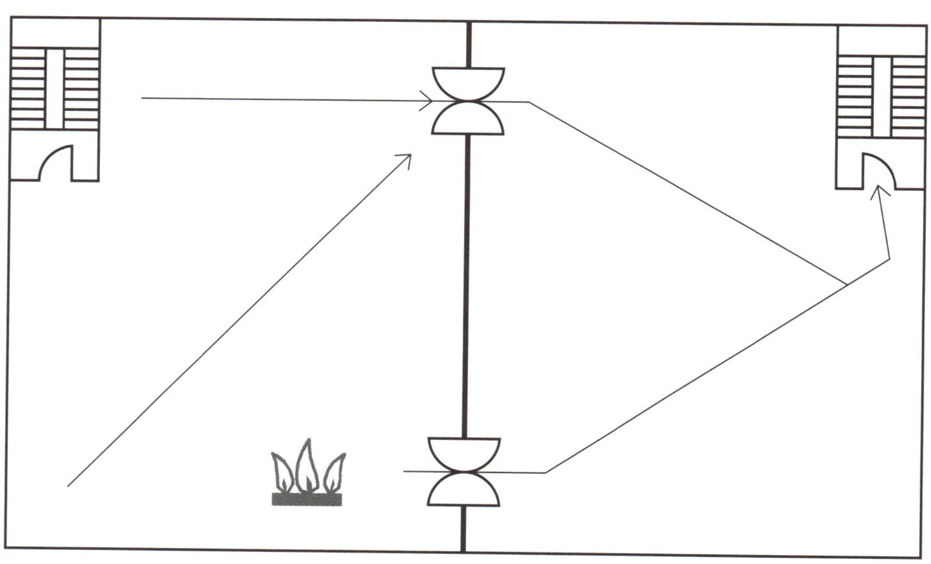

Storey divided into two refuges by compartment wall (stairways not provided with wheelchair space).

Note: Persons occupying the left-hand compartment would not reach a refuge until they had entered the right-hand compartment. Two doorsets in the partition are necessary in case access to one of the doorsets is blocked by fire.

4.10 Refuges and evacuation lifts should be clearly identified by appropriate fire safety signs. Where a refuge is in a lobby or stairway the sign should be accompanied by a blue mandatory sign worded "Refuge – keep clear".

Communication

4.11 To facilitate the effective evacuation of people from refuges an emergency voice communication (EVC) system should be provided. It is essential that the occupants of each refuge are able to alert other people that they are in need of assistance and for them to be reassured that this assistance will be forthcoming.

4.12 The EVC system should comply with BS 5839-9:2003 and consist of Type B outstations which communicate with a master station located in the building control room (where one exists) or adjacent to the fire alarm panel.

4.13 In some buildings it may be more appropriate to use an alternative approach such as the use of wireless technology.

Evacuation lifts

4.14 Guidance on the use of lifts when there is a fire is given in paragraph 5.39.

Width of escape stairs

4.15 The width of escape stairs should:

a. be not less than the width(s) required for any exit(s) affording access to them;

b. conform with the minimum widths given in Table 6;

c. not exceed 1400mm if their vertical extent is more than 30m, unless it is provided with a central handrail (see Notes 1 and 2); and

d. not reduce in width at any point on the way to a final exit.

Note 1: The 1400mm width has been given for stairs in tall buildings because research indicates that people prefer to stay within reach of a handrail when making a prolonged descent, so much so that the centre part of a wider stair is little used and could be hazardous. Thus additional stair(s) may be needed.

Note 2: Where a stair wider than 1400mm is provided with a central handrail, then the stair width on each side of the central handrail needs to be considered separately for the purpose of assessing stair capacity.

4.16 If the resultant width of the stair is more than 1800mm, then for reasons of safety in use the guidance in Approved Document K *Protection from falling, collision and impact* is that, in public buildings, the stair should have a central handrail. In such a case see Note 2 to paragraph 4.15.

4.17 Where an exit route from a stair also forms the escape from the ground and/or basement storeys, the width may need to be increased accordingly. (See paragraph 3.23).

Calculation of minimum stair width

General

4.18 Every escape stair should be wide enough to accommodate the number of persons needing to use it in an emergency. This width will depend on the number of stairs provided and whether the escape strategy for the building (or part of the building) is based on simultaneous evacuation see (paragraph 4.22) or phased evacuation see (paragraph 4.26).

4.19 As with the design of horizontal escape routes, where the maximum number of people needing to use the escape stairs is not known, the occupant capacity should be calculated on the basis of the appropriate floor space factors. Guidance for this is set out in Appendix C.

Table 6 Minimum widths of escape stairs

Situation of stair		Maximum number of of people served [1]	Minimum stair width (mm)
1a.	In an institutional building (unless the stair will only be used by staff)	150	1000 [2]
1b.	In an assembly building and serving an area used for assembly purposes (unless the area is less than 100m²)	220	1100
1c.	In any other building and serving an area with an occupancy of more than 50	Over 220	See Note [3]
2.	Any stair not described above	50	800 [4]

Notes:

1. Assessed as likely to use the stair in a fire emergency.
2. BS 5588-5 recommends that firefighting stairs should be at least 1100mm wide.
3. See Table 7 for sizing stairs for simultaneous evacuation, and Table 8 for phased evacuation.
4. In order to comply with the guidance in the Approved Document to Part M on minimum widths for areas accessible to disabled people, this may need to be increased to 1000mm.

Discounting of stairs

4.20 Whether phased or simultaneous evacuation is used, where two or more stairs are provided it should be assumed that one of them might not be available due to fire. It is therefore necessary to discount each stair in turn in order to ensure that the capacity of the remaining stair(s) is adequate for the number of persons needing to escape. The stair discounting rule applies to a building fitted with a sprinkler system.

4.21 Two exceptions to the above discounting rules are if the escape stairs:

a. are protected by a smoke control system designed in accordance with BS EN 12101-6:2005.

b. are approached on each storey through a protected lobby (a protected lobby need not be provided on the topmost storey for the exception still to apply).

Note: Paragraph 4.34 identifies several cases where stairs need lobby protection.

In such cases the likelihood of a stair not being available is significantly reduced and it is not necessary to discount a stair. However, a storey exit still needs to be discounted, see paragraph 3.21. See also paragraph 4.27 for additional guidance on the potential need to discount stairs in tall buildings utilising phased evacuation.

Simultaneous evacuation

4.22 In a building designed for simultaneous evacuation, the escape stairs (in conjunction with the rest of the means of escape) should have the capacity to allow all floors to be evacuated simultaneously. In calculating the width of the stairs account is taken of the number of people temporarily housed in the stairways during the evacuation.

4.23 Escape based on simultaneous evacuation should be used for:

a. all stairs serving basements;

b. all stairs serving buildings with open spatial planning; and

c. all stairs serving Other Residential or Assembly and Recreation buildings.

Note: BS 5588-7:1997 includes designs based on simultaneous evacuation.

Table 7 Capacity of a stair for basements and for simultaneous evacuation of the building

No. of floors served	Maximum number of persons served by a stair of width:								
	1000mm	1100mm	1200mm	1300mm	1400mm	1500mm	1600mm	1700mm	1800mm
1.	150	220	240	260	280	300	320	340	360
2.	190	260	285	310	335	360	385	410	435
3.	230	300	330	360	390	420	450	480	510
4.	270	340	375	410	445	480	515	550	585
5.	310	380	420	460	500	540	580	620	660
6.	350	420	465	510	555	600	645	690	735
7.	390	460	510	560	610	660	710	760	810
8.	430	500	555	610	665	720	775	830	885
9.	470	540	600	660	720	780	840	900	960
10.	510	580	645	710	775	840	905	970	1035

Notes:

1. The capacity of stairs serving more than 10 storeys may be obtained by using linear extrapolation.
2. The capacity of stairs not less than 1100mm wide may also be obtained by using the formula in paragraph 4.25.
3. Stairs with a rise of more than 30m should not be wider than 1400mm unless provided with a central handrail (see paragraph 4.15).
4. Stairs wider than 1800mm should be provided with a central handrail (see paragraph 4.16).

4.24 Where simultaneous evacuation is to be used, the capacity of stairs of widths from 1000 to 1800mm is given in Table 7.

4.25 As an alternative to using Table 7, the capacity of stairs 1100mm or wider (for simultaneous evacuation) can be derived from the formula:

$$P = 200w + 50\,(w - 0.3)(n - 1), \text{ or}$$

$$w = \frac{P + 15n - 15}{150 + 50n}$$

where:

(P) is the number of people that can be served; (w) is the width of the stair, in metres; and (n) is the number of storeys served.

Note 1: Stairs with a rise of more than 30m should not be wider than 1400mm unless provided with a central handrail (see paragraph 4.15).

Note 2: Separate calculations should be made for stairs/flights serving basement storeys and those serving upper storeys.

Note 3: The population 'P' should be divided by the number of available stairs.

Note 4: The formula is particularly useful when determining the width of stairs serving a building (or part of a building) where the occupants are not distributed evenly – either within a storey or between storeys.

Note 5: In the formula, the first part [200w] represents the number of persons estimated to have left the stair after 2.5 minutes of evacuation. The second part [50(w-0.3)(n-1)] represents the number of persons estimated to be accommodated on the stair after this time.

Worked examples:

A 14-storey building comprises 12 storeys of offices (ground + 11) with the top two storeys containing flats served by separate stairs. What is the minimum width needed for the stairs serving the office floors with a population of 1200 people (excluding the ground floor population which does not use the stairs), using simultaneous evacuation? Two stairs satisfy the travel distance limitations.

a. The population is distributed evenly.

As the top office storey is at a height greater than 18m, both stairs need lobby protection (see paragraph 4.34). Therefore, as both stairs are entered at each level via a protected lobby, then both stairs can be assumed to be available (see paragraph 4.21).

P = 1200/2 = 600, n = 11

From the formula:

600 = 200w + 50 (w – 0.3)(11 – 1)

600 = 200w + (50w – 15)(10)

600 = 200w + 500w – 150

750 = 700w

w = 1070mm

Therefore both stairs should be at least 1070mm wide. But this needs to be increased to 1100mm as the formula applies to stairs 1100mm or wider (see paragraph 4.25).

This width will also be adequate when one storey exit is discounted in accordance with paragraph 3.21 and the need to comply with paragraph 4.15(a) (i.e. the stair widths are not less than the minimum widths needed for 110 persons in Table 4).

b. The population is not distributed evenly

(e.g. 1000 people occupy floors 1 to 9 and 200 occupy floors 10 and 11).

As the top office storey is at a height greater than 18m, both stairs need lobby protection (see paragraph 4.34). As both stairs are entered at each level via a protected lobby, then both stairs can be assumed to be available (see paragraph 4.21).

To find the width of

- *the stairs serving floors 10 and 11:*

P = 200/2 = 100, n = 2

From the formula:

100 = 200w + 50 (w – 0.3)(2 – 1)

100 = 200w + (50w – 15)(1)

100 = 200w + 50w – 15

115 = 250w

w = 460mm

Therefore both stairs between the 9th floor landing and the top floor should be at least 460mm. But this needs to be increased to 1100mm as the formula applies to stairs 1100mm or wider (see paragraph 4.25).

This width will also be adequate when one storey exit is discounted in accordance with paragraph 3.21 and the need to comply with paragraph 4.15(a) (i.e. the stair widths are not less than the minimum widths needed for 100 persons in Table 4).

- *the stairs serving floors 1 to 9:*

P = 1200/2 = 600, n = 9

From the formula:

600 = 200w + 50 (w – 0.3) (9 – 1)

600 = 200w + (50w – 15) (8)

600 = 200w + 400w – 120

720 = 600w

w = 1200mm

Therefore both stairs between the 9th floor landing and the ground floor should be at least 1200mm wide.

This width will also be adequate when one storey exit is discounted in accordance with paragraph 2.21 and the need to comply with paragraph 4.15(a) (i.e. the stair widths are not less than the minimum widths needed for 111 persons in Table 4).

Phased evacuation

4.26 Where it is appropriate to do so, it may be advantageous to design stairs in high buildings on the basis of phased evacuation. In phased evacuation the first people to be evacuated are all those of reduced mobility and those on the storey most immediately affected by the fire. Subsequently, if there is a need to evacuate more people, it is done two floors at a time. It is a method which cannot be used in every type of building and it depends on the provision (and maintenance) of certain supporting facilities such as fire alarms. It does however enable narrower stairs to be incorporated than would be the case if simultaneous evacuation were used and has the practical advantage of reducing disruption in large buildings.

4.27 In tall buildings over 30m in height, where phased evacuation is adopted, there is a potential that persons attempting to escape could be impeded by firefighters entering and operating within the building. This potential varies with the height of the building and with the number of escape stairs that are available. Generally, this can be addressed by incorporating special management procedures into the evacuation strategy in consultation with Fire and Rescue Service. However, in some very tall buildings, typically those over 45m in height, physical measures may need to be incorporated into the building (e.g. by discounting a stair or by some other suitable means).

4.28 Phased evacuation may be used for any building provided it is not identified in paragraph 4.23 as needing simultaneous evacuation.

4.29 The following criteria should be satisfied in a building (or part of a building) that is designed on the basis of phased evacuation:

a. the stairways should be approached through a protected lobby or protected corridor at each storey, except a top storey;

b. the lifts should be approached through a protected lobby at each storey (see paragraph 5.42);

c. every floor should be a compartment floor;

d. if the building has a storey with a floor over 30m above ground level, the building should be protected throughout by an automatic sprinkler system in accordance with paragraph 0.16.

e. the building should be fitted with an appropriate fire warning system, conforming to at least the L3 standard given in BS 5839-1:2002; and

f. an internal speech communication system should be provided to permit conversation between a control point at fire and rescue service access level and a fire warden on every storey. In addition, the recommendations relating to phased evacuation provided in BS 5839-1 should be followed. Where it is deemed appropriate to install a voice alarm, this should be in accordance with BS 5839-8:1998.

4.30 The minimum width of stair needed when phased evacuation is used is given in Table 8. This table assumes a phased evacuation of the fire floor first followed by evacuation of not more than two floors at a time.

Table 8 Minimum width of stairs designed for phased evacuation

Maximum number of people in any storey	Stair width mm [1]
100	1000
120	1100
130	1200
140	1300
150	1400
160	1500
170	1600
180	1700
190	1800

Notes:

1. Stairs with a rise of more than 30m should not be wider than 1400mm unless provided with a central handrail (see paragraph 5.6).
2. As an alternative to using this table, provided that the minimum width of a stair is at least 1000mm, the width may be calculated from: [(P x 10) – 100]mm where P = the number of people on the most heavily occupied storey.

Worked example using Table 8

What is the minimum width needed for the stairs serving an 15-storey office building (ground + 14 office floors) assuming a total population of 2500 people (excluding the ground floor population which does not use the stairs). Three stairs satisfy the travel distance limitations.

The building is over 45 metres in height and designed for phased evacuation. It has been decided to discount one stair to take account of fire and rescue service operations as described in paragraphs 4.27. Therefore:

- *Number of persons per storey = 2500/14 = 179;*

Each remaining stair must be able to accommodate half the population of one storey (i.e. 90 persons)

Thus each stair requires a width of 1000mm (maximum capacity 100 persons)

This width will also be adequate when one storey exit is discounted in accordance with paragraph 3.21 and the need to comply with paragraph 4.15(a) (i.e. the stair widths are not less than the minimum width needed for 90 persons in Table 4).

- *At least one of those stairs will need to be a firefighting stair thus a minimum width of 1100mm wil be needed (see note 2 to Table 7).*

Additional worked example using Table 8

What is the minimum width needed for the stairs serving a 9-storey office building (ground + 8 office floors) assuming a total population of 1920 people (excluding the ground floor population which does not use the stairs). Two stairs satisfy the travel distance limitations.

As both stairs need to be entered at each level by a protected lobby (see paragraph 4.29), then both stairs can be assumed to be available (see paragraph 4.21). Therefore:

- *Number of persons per storey = 1920/8 = 240;*
- *Each stair must be able to accommodate half the population of one storey (i.e. 240/2 = 120 persons)*
- *Thus both stairs would require a width of 1100mm (maximum capacity 120 persons) according to Table 8, but:*
- *Each storey exit needs to be able to serve 240 persons due to discounting, in accordance with paragraph 3.21. The minimum exit width needed for 240 persons in Table 4 is 1200mm. In accordance with paragraph 4.15(a) the stair width should be at least as wide as the storey exit serving it.*
- *The required stair width is therefore 1200mm.*

Protection of escape stairs

General

4.31 Escape stairs need to have a satisfactory standard of fire protection if they are to fulfil their role as areas of relative safety during a fire evacuation. The guidance in paragraphs 4.32 to paragraph 4.33 should be followed to achieve this.

Enclosure of escape stairs

4.32 Every internal escape stair should be a protected stairway (i.e. it should be within a fire-resisting enclosure).

However an unprotected stair (e.g. an accommodation stair) may form part of an internal route to a storey exit or final exit, provided that the distance of travel and the number of people involved are very limited. For example, small premises (described in paragraph 3.32, 4.6 and 4.33) and raised storage areas (see paragraphs 7.7 and 7.8).

There may be additional measures if the protected stairway is also a protected shaft (where it penetrates one or more compartment floors, see Section 8) or if it is a firefighting shaft (see Section 17).

Small Premises

4.33 A stair in a small premises, which is not a bar or restaurant, may be open if it does not connect more than two storeys and delivers into the ground storey not more than 3m from the final exit (see Diagrams 22 and 23) and either:

a. the storey is also served by a protected stairway; or

b. it is a single stair in a small premises with the floor area in any storey not exceeding 90m^2 and, if the premises contains three storeys, the stair serving either the top or bottom storey is enclosed with fire-resisting construction at the ground storey level and discharges to a final exit independent of the ground storey (see Diagram 23).

Diagram 22 Maximum travel distance in a small two-storey premises with a single open stair

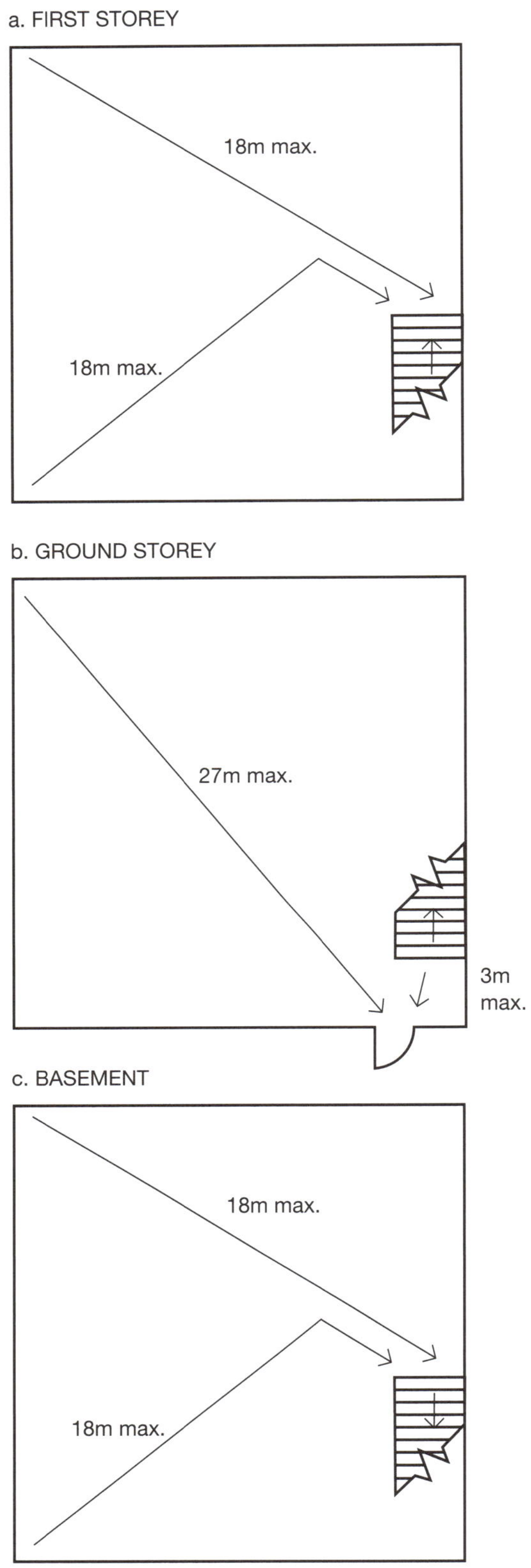

Note 1: Maximum floor area in any one storey $90m^2$.
Note 2: The premises may not be used as a restaurant or bar.
Note 3: Only acceptable in two storey premises (a+b or b+c).
Note 4: Travel distances are set out in Table 4.

Diagram 23 Maximum travel distance in a small three-storey premises with a single stair to each storey

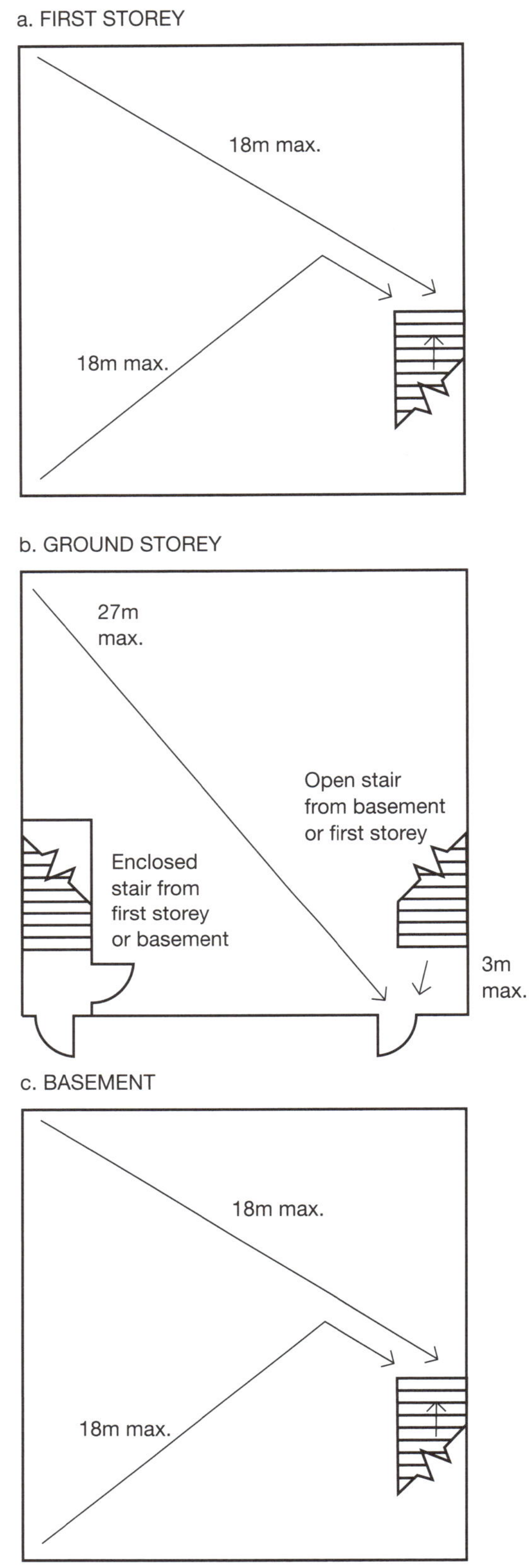

Note 1: Maximum floor area in any one storey $90m^2$.
Note 2: Enclosed stair at ground storey level may be from either the basement or the first storey.
Note 3: The premises may not be used as a restaurant or bar.
Note 4: Travel distances are set out in Table 4.

Access lobbies and corridors

4.34 There are situations where an escape stair needs the added protection of a protected lobby or protected corridor. These are:

a. where the stair is the only one serving a building (or part of a building) which has more than one storey above or below the ground storey (except for small premises covered in paragraph 4.6a); or

b. where the stair serves any storey at a height greater than 18m; or

c. where the building is designed for phased evacuation (see paragraph 4.29a).

In these cases protected lobbies or protected corridors are needed at all levels, except the top storey and at all basement levels; or

d. where the stair is a firefighting stair.

Lobbies are also needed where the option in paragraph 4.21(b) has been used so as not to discount one stairway when calculating stair widths.

An alternative that may be considered in (a) to (c) above is to use a smoke control system as described in paragraph 4.21(a).

4.35 A protected lobby should be also provided between an escape stairway and a place of special fire hazard. In this case, the lobby should have not less than 0.4m^2 permanent ventilation, or be protected from the ingress of smoke by a mechanical smoke control system.

Exits from protected stairways

4.36 Every protected stairway should discharge:

a. directly to a final exit; or

b. by way of a protected exit passageway to a final exit.

Note: Any such protected exit passageway should have the same standard of fire resistance and lobby protection as the stairway it serves.

The exit from a protected stairway should meet the provisions in paragraphs 5.30 to 5.34.

Separation of adjoining stairways

4.37 Where two protected stairways are adjacent, they and any protected exit passageways linking them to final exits, should be separated by an imperforate enclosure.

Use of space within protected stairways

4.38 A protected stairway needs to be free of potential sources of fire. Consequently, facilities that may be incorporated in a protected stairway are limited to the following:

a. sanitary accommodation or washrooms, so long as the accommodation is not used as a cloakroom. A gas water heater or sanitary towel incinerator may be installed in the accommodation but not any other gas appliance;

b. a lift well may be included in a protected stairway, if it is not a firefighting stair;

c. a reception desk or enquiry office area at ground or access level, if it is not in the only stair serving the building or part of the building. The reception or enquiry office area should not be more than 10m^2 in area; and/or

d. cupboards enclosed with fire-resisting construction, if it is not in the only stair serving the building or part of the building.

External walls of protected stairways

4.39 The external enclosures to protected stairways should meet the provisions in paragraph 5.24.

Gas service pipes in protected stairways

4.40 Gas service and installation pipes or associated meters should not be incorporated within a protected stairway unless the gas installation is in accordance with the requirements for installation and connection set out in the *Pipelines Safety Regulations 1996,* SI 1996 No 825 and the *Gas Safety (Installation and Use) Regulations 1998* SI 1998 No 2451. (See also paragraph 8.40.)

Basement stairs

4.41 Because of their situation, basement stairways are more likely to be filled with smoke and heat than stairs in ground and upper storeys.

Special measures are therefore needed in order to prevent a basement fire endangering upper storeys. These are set out in the following two paragraphs.

4.42 If an escape stair forms part of the only escape route from an upper storey of a building (or part of a building) it should not be continued down to serve any basement storey. The basement should be served by a separate stair.

4.43 If there is more than one escape stair from an upper storey of a building (or part of a building), only one of the stairs serving the upper storeys of the building (or part) need be terminated at ground level. Other stairs may connect with the basement storey(s) if there is a protected lobby, or a protected corridor between the stair(s) and accommodation at each basement level.

External escape stairs

4.44 If more than one escape route is available from a storey (or part of a building), some of the escape routes from that storey or part of the building may be by way of an external escape stair, provided that:

a. there is at least one internal escape stair from every part of each storey (excluding plant areas);

b. in the case of an Assembly and Recreation building, the route is not intended for use by members of the public; or

c. in the case of an Institutional building, the route serves only office or residential staff accommodation.

4.45 Where external stairs are acceptable as forming part of an escape route, they should meet the provisions in paragraph 5.25.

Section 5: General provisions

Introduction

5.1 This Section gives guidance on the construction and protection of escape routes generally, service installations and other matters associated with the design of escape routes. It applies to all buildings.

It should therefore be read in conjunction with Section 2 (in respect of flats) and in conjunction with Sections 3 and 4 (in respect of other buildings).

Protection of escape routes

Fire resistance of enclosures

5.2 Details of fire resistance test criteria and standards of performance, are set out in Appendix A. Generally, a 30-minute standard is sufficient for the protection of means of escape. The exceptions to this are when greater fire resistance is required by the guidance on Requirements B3 or B5, or some other specific instance to meet Requirement B1, in Sections 2 and 3.

5.3 All walls, partitions and other enclosures that need to be fire-resisting to meet the provisions in this Approved Document (including roofs that form part of a means of escape), should have the appropriate performance given in Tables A1 and A2 of Appendix A.

5.4 Elements protecting a means of escape should meet any limitations on the use of glass (see paragraph 5.7).

Fire resistance of doors

5.5 Details of fire resistance test criteria and standards of performance, are set out in Appendix B.

5.6 All doors that need to be fire-resisting to meet the provisions in this Approved Document should have the appropriate performance given in Table B1 of Appendix B.

Doors should also meet any limitations on the use of glass (see paragraph 5.7).

Fire resistance of glazed elements

5.7 Where glazed elements in fire-resisting enclosures and doors are only able to satisfy the relevant performance in terms of integrity, the use of glass is limited. These limitations depend on whether the enclosure forms part of a protected shaft (see Section 8) and the provisions set out in Appendix A, Table A4.

5.8 Where the relevant performance can be met in terms of both integrity and insulation, there is no restriction in this Approved Document on the use or amount of glass.

5.9 Attention is also drawn to the guidance on the safety of glazing in Approved Document N *Glazing – safety in relation to impact, opening and cleaning.*

Doors on escape routes

5.10 The time taken to negotiate a closed door can be critical in escaping. Doors on escape routes (both within and from the building) should therefore be readily openable, if undue delay is to be avoided. Accordingly the provisions in paragraphs 5.11 to 5.18 should be met.

Door fastenings

5.11 In general, doors on escape routes (whether or not the doors are fire doors), should either not be fitted with lock, latch or bolt fastenings, or they should only be fitted with simple fastenings that can be readily operated from the side approached by people making an escape. The operation of these fastenings should be readily apparent; without the use of a key and without having to manipulate more than one mechanism. This is not intended to prevent doors being fitted with hardware to allow them to be locked when the rooms are empty. There may also be situations such as hotel bedrooms where locks may be fitted that are operated from the outside by a key and from the inside by a knob or lever, etc.

Where a door on an escape route has to be secured against entry when the building or part of the building is occupied, it should only be fitted with a lock or fastening which is readily operated, without a key, from the side approached by people making their escape. Similarly, where a secure door is operated by a code, combination, swipe or proximity card, biometric data or similar means, it should also be capable of being overridden from the side approached by people making their escape.

Electrically powered locks should return to the unlocked position:

a. on operation of the fire alarm system;

b. on loss of power or system error;

c. on activation of a manual door release unit (Type A) to BS EN 54-11:2001 positioned at the door on the side approached by people making their escape. Where the door provides escape in either direction, a unit should be installed on both sides of the door.

5.12 In the case of places of assembly, shop and commercial buildings, doors on escape routes from rooms with an occupant capacity of more than 60 should either not be fitted with lock, latch or bolt fastenings, or be fitted with panic fastenings in accordance with BS EN 1125:1997.

In non-residential buildings it may also be appropriate to accept on some final exit doors locks for security that are used only when the building is empty. In these cases the emphasis for the safe use of these locks must be placed on management procedures.

5.13 Guidance about door closing and 'hold open' devices for fire doors is given in Appendix B.

Direction of opening

5.14 The door of any doorway or exit should, if reasonably practicable, be hung to open in the direction of escape and should always do so if the number of persons that might be expected to use the door at the time of a fire is more than 60.

Note: Where there is a very high fire risk with potential for rapid fire growth, such as with some industrial activities, doors should open in the direction of escape even where the number of persons does not exceed 60.

Amount of opening and effect on associated escape routes

5.15 All doors on escape routes should be hung to open not less than 90 degrees with a swing that is clear of any change of floor level, other than a threshold or single step on the line of the doorway (see paragraph 5.21) and which does not reduce the effective width of any escape route across a landing.

5.16 A door that opens towards a corridor or a stairway should be sufficiently recessed to prevent its swing from encroaching on the effective width of the stairway or corridor.

Vision panels in doors

5.17 Vision panels are needed where doors on escape routes sub-divide corridors, or where any doors are hung to swing both ways. Note also the provision in Approved Document M *Access to and Use of buildings,* concerning vision panels in doors across accessible corridors and passageways and the provisions for the safety of glazing in Approved Document N *Glazing – safety in relation to impact, opening and cleaning.*

Revolving and automatic doors

5.18 Revolving doors, automatic doors and turnstiles can obstruct the passage of persons escaping. Accordingly, they should not be placed across escape routes unless:

a. they are to the required width and are automatic doors and either they:

   i. are arranged to fail safely to outward opening from any position of opening; or

   ii. are provided with a monitored failsafe system for opening the doors if the mains supply fails; or

   iii. they fail safely to the open position in the event of power failure; or

b. non-automatic swing doors of the required width are provided immediately adjacent to the revolving or automatic door or turnstile.

Stairs

Construction of escape stairs

5.19 The flights and landings of every escape stair should be constructed of materials of limited combustibility in the following situations:

a. if it is the only stair serving the building, or part of the building, unless the building is of two or three storeys and is in Purpose Group 1(a) or Purpose Group 3;

b. if it is within a basement storey (this does not apply to a private stair in a flat);

c. if it serves any storey having a floor level more than 18m above ground or access level;

d. if it is external, except in the case of a stair that connects the ground floor or paving level with a floor or flat roof not more than 6m above or below ground level. (There is further guidance on external escape stairs in paragraph 5.25); or

e. if it is a firefighting stair (see Section 17).

Note: In satisfying the above conditions, combustible materials may be added to the horizontal surface of these stairs (except in the case of firefighting stairs).

5.20 There is further guidance on the construction of firefighting stairs in Section 17. Dimensional constraints on the design of stairs generally, to meet requirements for safety in use, are given in Approved Document K, *Protection from falling, collision and impact.*

Single steps

5.21 Single steps may cause falls and should only be used on escape routes where they are prominently marked. A single step on the line of a doorway is acceptable, subject to paragraph 5.32.

Helical stairs, spiral stairs and fixed ladders

5.22 Helical stairs, spiral stairs and fixed ladders may form part of an escape route subject to the following restrictions:

a. helical and spiral stairs should be designed in accordance with BS 5395-2:1984 and, if they are intended to serve members of the public, should be a type E (public) stair, in accordance with that standard; and

b. fixed ladders should not be used as a means of escape for members of the public and should only be intended for use in circumstances where it is not practical to provide a conventional stair, for example, as access to plant rooms that are not normally occupied.

5.23 Guidance on the design of helical and spiral stairs and fixed ladders, from the aspect of safety in use, is given in Approved Document K *Protection from falling, collision and impact.*

External walls of protected stairways

5.24 With some configurations of external wall, a fire in one part of a building could subject the external wall of a protected stairway to heat (for example, where the two are adjacent at an internal angle in the facade as shown in Diagram 24). If the external wall of the protected stairway has little fire resistance, there is a risk that this could prevent the safe use of the stair.

Therefore, if:

a. a protected stairway projects beyond, or is recessed from, or is in an internal angle of, the adjoining external wall of the building; then

b. the distance between any unprotected area in the external enclosures to the building and any unprotected area in the enclosure to the stairway should be at least 1800mm (see Diagram 24).

External escape stairs

5.25 Where an external escape stair is provided in accordance with paragraph 4.44, it should meet the following provisions:

a. all doors giving access to the stair should be fire-resisting and self-closing, except that a fire-resisting door is not required at the head of any stair leading downwards where there is only one exit from the building onto the top landing;

b. any part of the external envelope of the building within 1800mm of (and 9m vertically below), the flights and landings of an external escape stair should be of fire-resisting construction, except that the 1800mm dimension may be reduced to 1100mm above the top level of the stair if it is not a stair up from a basement to ground level (see Diagram 25);

c. there is protection by fire-resisting construction for any part of the building (including any doors) within 1800mm of the escape route from the stair to a place of safety, unless there is a choice of routes from the foot of the stair that would enable the people escaping to avoid exposure to the effects of the fire in the adjoining building;

d. any stair more than 6m in vertical extent is protected from the effects of adverse weather conditions. (This should not be taken to imply a full enclosure. Much will depend on the location of the stair and the degree of protection given to the stair by the building itself); and

e. glazing in areas of fire-resisting construction mentioned above should also be fire-resisting (integrity but not insulation) and fixed shut.

Diagram 24 External protection to protected stairways

See para 5.24

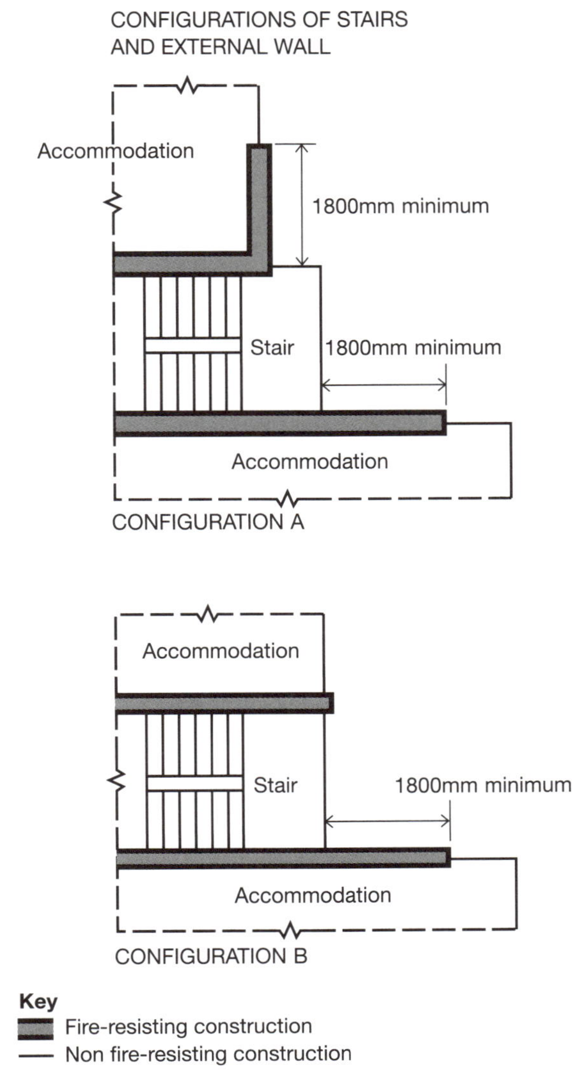

Diagram 25 **Fire resistance of areas adjacent to external stairs**

See para 5.25

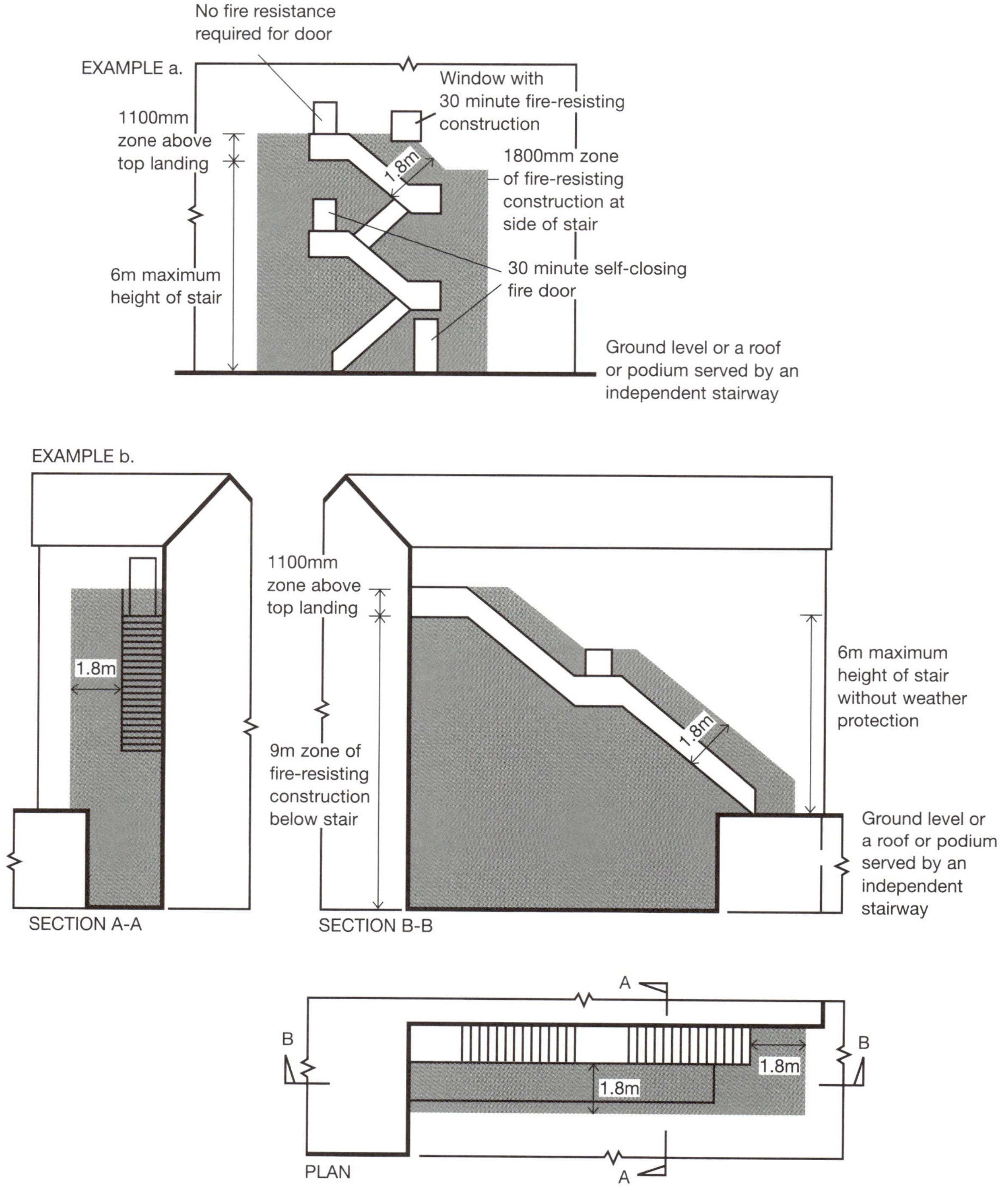

General

Headroom in escape routes

5.26 All escape routes should have a clear headroom of not less than 2m and there should be no projection below this height (except for door frames).

Floors of escape routes

5.27 The floorings of all escape routes (including the treads of steps and surfaces of ramps and landings) should be chosen to minimise their slipperiness when wet.

Ramps and sloping floors

5.28 Where a ramp forms part of an escape route it should meet the provisions in Approved Document M *Access to and Use of buildings.* Any sloping floor or tier should be constructed with a pitch of not more than 35° to the horizontal.

5.29 Further guidance on the design of ramps and associated landings and on aisles and gangways in places where there is fixed seating, from the aspect of safety in use, is given in Approved Document K *Protection from falling, collision and impact and* in Approved Document M *Access to and Use of buildings* . The design of means of escape in places with fixed seating is dealt with in Section 3 by reference to BS 5588-6:1991.

Final exits

5.30 Final exits need to be dimensioned and sited to facilitate the evacuation of persons out of and away from the building. Accordingly, they should be not less in width than the minimum width required for the escape route(s) they serve and should also meet the conditions in paragraphs 5.31 to 5.34.

5.31 Final exits should be sited to ensure rapid dispersal of persons from the vicinity of the building so that they are no longer in danger from fire and smoke. Direct access to a street, passageway, walkway or open space should be available. The route clear of the building should be well defined and, if necessary, have suitable guarding.

5.32 Final exits should not present an obstacle to wheelchair users and other people with disabilities. Where a final exit is accessed without the need to first traverse steps then a level threshold and, where necessary, a ramp should be provided.

5.33 Final exits need to be apparent to persons who may need to use them. This is particularly important where the exit opens off a stair that continues down, or up, beyond the level of the final exit.

5.34 Final exits should be sited so that they are clear of any risk from fire or smoke in a basement (such as the outlets to basement smoke vents, see Section 18), or from openings to transformer chambers, refuse chambers, boiler rooms and similar risks.

Escape routes over flat roofs

5.35 Where an escape route over a flat roof is provided in accordance with paragraph 2.31 or 3.31, it should meet the following provisions:

a. the roof should be part of the same building from which escape is being made;

b. the route across the roof should lead to a storey exit or external escape route;

c. the part of the roof forming the escape route and its supporting structure, together with any opening within 3m of the escape route, should be fire-resisting (see Appendix A Table A1); and

d. the route should be adequately defined and guarded by walls and/or protective barriers which meet the provisions in Approved Document K, *Protection from falling, collision and impact.*

Lighting of escape routes

5.36 All escape routes should have adequate artificial lighting. Routes and areas listed in Table 9 should also have escape lighting which illuminates the route if the main supply fails.

Lighting to escape stairs should be on a separate circuit from that supplying any other part of the escape route.

Standards for the installation of a system of escape lighting are given in BS 5266-1:2005.

Exit signs

5.37 Except within a flat, every escape route (other than those in ordinary use) should be distinctively and conspicuously marked by emergency exit sign(s) of adequate size complying with the *Health and Safety (Safety signs and signals) Regulations 1996.* In general, signs containing symbols or pictograms which conform to BS 5499-1:2002, satisfy these regulations. In some buildings additional signs may be needed to meet requirements under other legislation.

Suitable signs should also be provided for refuges (see paragraph 4.10).

Note: Advice on fire safety signs, including emergency escape signs, is given in an HSE publication: *Safety Signs and Signals: Guidance on Regulations.*

Protected power circuits

5.38 Where it is critical for electrical circuits to be able to continue to function during a fire, protected circuits are needed. The potential for damage to cables forming protected circuits should be limited by the use of sufficiently robust cables, careful selection of cable routes and/or by the provision of physical protection in areas where cables may be susceptible to damage. Methods of cable support should generally be non-combustible and such that circuit integrity will not be reduced below that afforded by the cable.

Table 9 **Provisions for escape lighting**

Purpose group of the building or part of the building	Areas requiring escape lighting
1. Residential	All common escape routes [1], except in 2-storey flats
2. Office, Storage and Other non-residential	a. Underground or windowless accommodation b. Stairways in a central core or serving storey(s) more than 18m above ground level c. Internal corridors more than 30m long d. Open-plan areas of more than $60m^2$
3. Shop and Commercial and car parks	a. Underground or windowless accommodation b. Stairways in a central core or serving storey(s) more than 18m above ground level c. Internal corridors more than 30m long d. Open-plan areas of more than $60m^2$ e. All escape routes to which the public are admitted [1] (except in shops of three or fewer storeys with no sales floor more than $280m^2$, provided that the shop is not a restaurant or bar)
4. Assembly and Recreation	All escape routes [1], and accommodation except for: a. accommodation open on one side to view sport or entertainment during normal daylight hours
5. Any Purpose Group	a. All toilet accommodation with a floor area over $8m^2$ b. Electricity and generator rooms c. Switch room/battery room for emergency lighting system d. Emergency control room

Notes:

1. Including external escape routes.

A protected circuit for operation of equipment in the event of fire should consist of cable meeting at least the requirements for PH 30 classification when tested in accordance with BS EN 50200:2006 (incorporating Appendix E), or an equivalent standard. It should follow a route selected to pass only through parts of the building in which the fire risk is negligible and should be separate from any circuit provided for another purpose.

In large or complex buildings there may be fire protection systems that need to operate for an extended period during a fire. Further guidance on the selection of cables for such systems is given in BS 5839-1, BS 5266-1 and BS 7346-6.

Lifts

Evacuation lifts

5.39 In general it is not appropriate to use lifts when there is a fire in the building because there is always the danger of people being trapped in a lift that has become immobilised as a result of the fire. However, in some circumstances a lift may be provided as part of a management plan for evacuating people. In such cases the lift installation may need to be appropriately sited and protected and may need to contain a number of safety features that are intended to ensure that the lift remains usable for evacuation purposes during the fire. Guidance on the design and use of evacuation lifts is given in BS 5588-*8:1999.*

Where a firefighting lift has been provided to satisfy requirement B5, this can be utilised as part of a management plan for evacuating disabled people. Any such plan should include a contingency for when the Fire and Rescue Service arrive.

Fire protection of lift installations

5.40 Because lifts connect floors, there is the possibility that they may prejudice escape routes. To safeguard against this, the conditions in paragraphs 5.41 to 5.45 should be met.

5.41 Lifts, such as wall-climber or feature lift which rise within a large volume, such as a mall or atrium, and do not have a conventional well, may be at risk if they run through a smoke reservoir. In which case, care is needed to maintain the integrity of the smoke reservoir and protect the occupants of the lift.

5.42 Lift wells should be either:

a. contained within the enclosures of a protected stairway; or

b. enclosed throughout their height with fire-resisting construction if they are sited so as to prejudice the means of escape.

A lift well connecting different compartments should form a protected shaft (see Section 8).

In buildings designed for phased or progressive horizontal evacuation, where the lift well is not contained within the enclosures of a protected stairway, the lift entrance should be separated from the floor area on every storey by a protected lobby.

5.43 In basements and enclosed (non open-sided) car parks the lift should be approached only by a protected lobby (or protected corridor), unless it is within the enclosure of a protected stairway.

This is also the case in any storey that contains high fire risk areas, if the lift also delivers directly into corridors serving sleeping accommodation. Examples of fire risk areas in this context are kitchens, communal lounges and stores.

5.44 A lift shaft should not be continued down to serve any basement storey if it is:

a. in a building (or part of a building) served by only one escape stair and smoke from a basement fire would be able to prejudice the escape routes in the upper storeys; or

b. within the enclosure to an escape stair which is terminated at ground level.

5.45 Lift machine rooms should be sited over the lift well whenever possible. If the lift well is within a protected stairway which is the only stairway serving the building (or part of the building), then if the machine room cannot be sited above the lift well, it should be located outside the stairway (to avoid smoke spread from a fire in the machine room).

Mechanical ventilation and air-conditioning systems

5.46 Any system of mechanical ventilation should be designed to ensure that, in a fire, the ductwork does not assist in transferring fire and smoke through the building and put at risk the protected means of escape from the accommodation areas. Any exhaust points should be sited so as not to further jeopardize the building, i.e. away from final exits, combustible building cladding or roofing materials and openings into the building.

5.47 Ventilation ducts supplying or extracting air directly to or from a protected escape route, should not also serve other areas. A separate ventilation system should be provided for each protected stairway. Guidance on ventilation systems that circulate air only within an individual flat is given in paragraph 2.18.

Where the ductwork system serves more than one part of a sub-divided (see paragraph 3.26) escape route, a fire damper should be provided where ductwork enters each section of the escape route operated by a smoke detector or suitable fire detection system (see also Section 10). The fire dampers should close when smoke is detected.

5.48 Ducts passing through the enclosure of a protected escape route should be fire-resisting, i.e. the ductwork should be constructed in accordance with Method 2 or Method 3, (see paragraph 10.9).

Note: Fire dampers activated only by fusible links are not suitable for protecting escape routes. However an ES classified fire and smoke damper which is activated by a suitable fire detection system may be used. See paragraph 10.15.

5.49 In the case of a system which recirculates air, smoke detectors should be fitted in the extract ductwork before the point of separation of the recirculated air and the air to be discharged to the open air and before any filters or other air cleaning equipment. Such detector(s) should:

a. cause the system to immediately shut down; or

b. switch the ventilation system from recirculating mode to extraction to open air, so as to divert the any smoke to the outside of the building.

5.50 Non-domestic kitchens, car parks and plant rooms should have separate and independent extraction systems and the extracted air should not be recirculated.

5.51 Guidance on the use of mechanical ventilation in a place of assembly is given in BS 5588-6:1991.

5.52 Where a pressure differential system is installed, ventilation and air-conditioning systems in the building should be compatible with it when operating under fire conditions.

5.53 Further guidance on the design and installation of mechanical ventilation and air conditioning plant is given in BS 5720:1979. Guidance on the provision of smoke detectors in ventilation ductwork is given in BS 5839-1:2002.

Note: Paragraphs 8.41 and 10.9 also deal with ventilation and air-conditioning ducts.

Refuse chutes and storage

5.54 Refuse storage chambers, refuse chutes and refuse hoppers should be sited and constructed in accordance with BS 5906 *Code of practice for storage and on-site treatment of solid waste from buildings.*

5.55 Refuse chutes and rooms provided for the storage of refuse should:

a. be separated from other parts of the building by fire-resisting construction; and

b. not be located within protected stairways or protected lobbies.

5.56 Rooms containing refuse chutes, or provided for the storage of refuse, should be approached either directly from the open air or by way of a protected lobby provided with not less than $0.2m^2$ of permanent ventilation.

5.57 Access to refuse storage chambers should not be sited adjacent to escape routes or final exits, or near to windows of flats.

Shop store rooms

5.58 Fully enclosed walk-in store rooms in shops (unless provided with an automatic fire detection and alarm system or fitted with sprinklers) should be separated from retail areas with fire-resisting construction (see Appendix A, Table A1), if they are sited so as to prejudice the means of escape.

The Requirement

This Approved Document deals with the following Requirement from Part B of Schedule 1 to the Building Regulations 2000 (as amended).

Requirement	*Limits on application*
Internal fire spread (linings) **B2.** (1) To inhibit the spread of fire within the building, the internal linings shall: (a) adequately resist the spread of flame over their surfaces; and (b) have, if ignited, a rate of heat release or a rate of fire growth, which is reasonable in the circumstances. (2) In this paragraph 'internal linings' mean the materials or products used in lining any partition, wall, ceiling or other internal structure.	

Insurers' Essential Principles

The relevant essential principles for property protection are:

- With the exception of joinery products, the building shall be constructed from building materials/products that will not make a significant contribution to the early stages of a fire or contribute to the spread of fire;
- The building shall be constructed in such a manner that if a fire starts, the extent of fire and smoke damage will be minimised and confined as close to the source of fire outbreak as is practical/feasible;
- There shall be adequate provision to prevent an arson attack.

Guidance

Performance

In the Secretary of State's view the Requirements of B2 will be met if the spread of flame over the internal linings of the building is restricted by making provision for them to have low rates of surface spread of flame and, in some cases, to have a low rate of heat release, so as to limit the contribution that the fabric of the building makes to fire growth. In relation to the European fire tests and classification system, the requirements of B2 will be met if the heat released from the internal linings is restricted by making provision for them to have a resistance to ignition and a rate of fire growth which are reasonable in the circumstances.

The extent to which this is necessary is dependent on the location of the lining.

Introduction

Fire spread and lining materials

B2.i The choice of materials for walls and ceilings can significantly affect the spread of a fire and its rate of growth, even though they are not likely to be the materials first ignited.

It is particularly important in circulation spaces where linings may offer the main means by which fire spreads and where rapid spread is most likely to prevent occupants from escaping.

Several properties of lining materials influence fire spread. These include the ease of ignition and the rate at which the lining material gives off heat when burning. The guidance relating to the European fire tests and classification provides for control of internal fire spread through control of these properties. This document does not give detailed guidance on other properties such as the generation of smoke and fumes.

Floors and stairs

B2.ii The provisions do not apply to the upper surfaces of floors and stairs because they are not significantly involved in a fire until well developed and thus do not play an important part in fire spread in the early stages of a fire that are most relevant to the safety of occupants.

However, it should be noted that the construction of some stairs and landings is controlled under Section 5, paragraph 5.19 and in the case of firefighting stairs, Section 17, paragraph 17.11.

Other controls on internal surface properties

B2.iii There is also guidance on the control of flame spread inside buildings in two other Sections. In Section 8 there is guidance on surfaces exposed in concealed spaces above fire-protecting suspended ceilings and in Section 10 on enclosures to above ground drainage system pipes.

Note: External flame spread is dealt with in Sections 12 to 14; the fire behaviour of insulating core panels used for internal structures is dealt with in Appendix F.

Furniture and fittings

B2.iv Furniture and fittings can have a major effect on fire spread but it is not possible to control them through Building Regulations and they are not dealt with in this Approved Document. Fire characteristics of furniture and fittings may be controlled in some buildings under legislation that applies to a building in use, such as licensing conditions.

From a property protection viewpoint, insurers will take into account the contribution to a developing fire that the entire contents of the building will make. Additionally, the financial exposure of the contents of the building will be a significant factor. This may lead to a recommendation that the provision of an automatic extinguishing system is considered.

Classification of performance

B2.v Appendix A describes the different classes of performance and the appropriate methods of test (see paragraphs 7 to 20).

The National classifications used are based on tests in BS 476 *Fire tests on building materials and structures,* namely Part 6: *Method of test for fire propagation for products* and Part 7: *Method of test to determine the classification of the surface spread of flame of products.* However, Part 4: *Non-combustibility test for materials* and Part 11: *Method for assessing the heat emission from building products* are also used as one method of meeting Class 0. Other tests are available for classification of thermoplastic materials if they do not have the appropriate rating under BS 476-7 and three ratings, referred to as TP(a) rigid and TP(a) flexible and TP(b), are used.

The European classifications are described in BS EN 13501-1:2002, *Fire classification of construction products and building elements,* Part 1 – *Classification using data from reaction to fire tests.* They are based on a combination of four European test methods, namely:

- BS EN ISO 1182:2002 *Reaction to fire tests for building products – Non combustibility test;*
- BS EN ISO 1716:2002 *Reaction to fire tests for building products – Determination of the gross calorific value;*
- BS EN 13823:2002 *Reaction to fire tests for building products – Building products excluding floorings exposed to the thermal attack by a single burning item;* and
- BS EN ISO 11925-2:2002, *Reaction to fire tests for building products. Ignitability when subjected to direct impingement of flame.*

For some building products, there is currently no generally accepted guidance on the appropriate procedure for testing and classification in accordance with the harmonised European fire tests. Until such a time that the appropriate European test and classification methods for these building products are published, classification may only be possible using existing national test methods.

Table A8, in Appendix A, gives typical performance ratings which may be achieved by some generic materials and products.

Section 6: Wall and ceiling linings

Classification of linings

6.1 Subject to the variations and specific provisions described in paragraphs 6.2 to 6.16, the surface linings of walls and ceilings should meet the following classifications:

Table 10 Classification of linings

Location	National class (1)	European class (1)(3)(4)
Small rooms (2) of area not more than: a. $4m^2$ in residential accommodation b. $30m^2$ in non-residential accommodation	3	D-s3, d2
Other rooms (2) (including garages)	1	C-s3, d2
Circulation spaces within dwellings		
Other circulation spaces, including the common areas of blocks of flats	0	B-s3, d2

Notes:

1. See paragraph B2.v.
2. For meaning of room, see definition in Appendix E.
3. The National classifications do not automatically equate with the equivalent classifications in the European column, therefore, products cannot typically assume a European class, unless they have been tested accordingly.
4. When a classification includes 's3, d2', this means that there is no limit set for smoke production and/or flaming droplets/particles.

See Table 10a for property protection recommendations.

Table 10a Classification of linings (property protection)

Location	National class (1)	European class (1)(3)(4)
Small rooms (2) of area not more than: $4m^2$ in residential accommodation	3	D-s3, d2
$30m^2$ in non-residential accommodation	1	C-s3, d2
Other rooms (including garages)		
Circulation spaces within dwellings		
Other circulation spaces, including the common areas of blocks of flats	0	B-s3, d2
Areas of special risk, including plant rooms, storage, data processing	Limited combustibility	A2, B-s1, d0

Notes:

1. See paragraph B2.v.
2. For meaning of room, see definition in Appendix E.
3. The National classifications do not automatically equate with the equivalent classifications in the European column, therefore, products cannot typically assume a European class, unless they have been tested accordingly.
4. When a classification includes 's3, d2', this means that there is no limit set for smoke production and/or flaming droplets/particles.

Table 10a is intended to relate to surface properties only. Ideally, substrates should be either:

- *non-combustible; or*
- *of limited combustibility; or*
- *Euro class A1 or A2-s1, d0 or B-s1, d0; or*
- *confirmed as having been tested to, and meeting in full, the requirements of LPS 1181 or another test/approval standard acceptable to insurers.*

Definition of walls

6.2 For the purpose of the performance of wall linings, a wall includes:

a. the surface of glazing (except glazing in doors); and

b. any part of a ceiling which slopes at an angle of more than 70° to the horizontal.

But a wall does not include:

c. doors and door frames;

d. window frames and frames in which glazing is fitted;

e. architraves, cover moulds, picture rails, skirtings and similar narrow members; or

f. fireplace surrounds, mantle shelves and fitted furniture.

Definition of ceilings

6.3 For the purposes of the performance of ceiling linings, a ceiling includes:

a. the surface of glazing;

b. any part of a wall which slopes at an angle of 70° or less to the horizontal;

c. the underside of a mezzanine or gallery; and

d. the underside of a roof exposed to the room below.

But a ceiling does not include:

e. trap doors and their frames;

f. the frames of windows or rooflights (see Appendix E) and frames in which glazing is fitted; or

g. architraves, cover moulds, picture rails, exposed beams and similar narrow members.

Variations and special provisions

Walls

6.4 Parts of walls in rooms may be of a poorer performance than specified in paragraph 6.1 and Table 10 (but not poorer than Class 3 (National class) or Class D-s3, d2 (European class)), provided the total area of those parts in any one room does not exceed one half of the floor area of the room; and subject to a maximum of 20m^2 in residential accommodation and 60m^2 in non-residential accommodation.

For purposes of property protection, the provisions in Table 10a still apply.

Fire-protecting suspended ceilings

6.5 A suspended ceiling can contribute to the overall fire resistance of a floor/ceiling assembly. Such a ceiling should satisfy paragraph 6.1 and Table 10. It should also meet the provisions of Appendix A, Table A3.

Fire-resisting ceilings

6.6 Cavity barriers are needed in some concealed floor or roof spaces (see Section 9); however, this need can be reduced by the use of a fire-resisting ceiling below the cavity. Such a ceiling should comply with Diagram 35.

Rooflights

6.7 Rooflights should meet the relevant classification in 6.1 and Table 10. However plastic rooflights with at least a Class 3 rating may be used where 6.1 calls for a higher standard, provided the limitations in Table 11 and Table 18 are observed.

For purposes of property protection, the provisions in Table 10a still apply. Where the building is protected by sprinklers, plastic rooflights with a softening temperature below 200°C should not be specified. For plastic rooflights used in areas that are ESFR (early suppression fast response) sprinkler protected, the *LPC Rules for automatic sprinkler installations*, TB209.9.4 require rooflights to survive an exposure to 300°C for five minutes.

Note: No guidance is currently possible on the performance requirements in the European fire tests as there is no generally accepted test and classification procedure.

Special applications

6.8 Any flexible membrane covering a structure (other than an air supported structure) should comply with the recommendations given in Appendix A of BS 7157:1989.

6.9 Guidance on the use of PTFE-based materials for tension-membrane roofs and structures is given in a BRE report *Fire safety of PTFE-based materials used in buildings* (BR 274, BRE 1994).

Thermoplastic materials

General

6.10 Thermoplastic materials (see Appendix A, paragraph 17) which cannot meet the performance given in Table 10, can nevertheless be used in windows, rooflights and lighting diffusers in suspended ceilings if they comply with the provisions described in paragraphs 6.11 to 6.15. Flexible thermoplastic material may be used in panels to form a suspended ceiling if it complies with the guidance in paragraph 6.16. The classifications used in paragraphs 6.11 to 6.16, Table 11 and Diagram 27 are explained in Appendix A, paragraph 20.

Note: No guidance is currently possible on the performance requirements in the European fire tests as there is no generally accepted test and classification procedure.

Windows and internal glazing

6.11 External windows to rooms (though not to circulation spaces) may be glazed with thermoplastic materials, if the material can be classified as a TP(a) rigid product.

Internal glazing should meet the provisions in paragraph 6.1 and Table 10 above.

Note 1: A "wall" does not include glazing in a door (see paragraph 6.2).

Note 2: Attention is drawn to the guidance on the safety of glazing in Approved Document N *Glazing – safety in relation to impact, opening and cleaning*.

Note 3: Where plastic window frames are used, cavity barriers around the opening should be specified to provide a minimum of 30 minutes fire resistance in terms of integrity and insulation.

It is increasingly common for combustible insulants to be located in cavities in external walls. The objective of the above recommendation is to prevent, as far as possible, ignition of combustibles in the cavity from a fire entering or exiting the window opening.

Rooflights

6.12 Rooflights to rooms and circulation spaces (with the exception of protected stairways or areas protected by sprinklers) may be constructed of a thermoplastic material if:

a. the lower surface has a TP(a) (rigid) or TP(b) classification;

b. the size and disposition of the rooflights accords with the limits in Table 11 and with the guidance to B4 in Tables 17 and 18.

Lighting diffusers

6.13 The following provisions apply to lighting diffusers which form part of a ceiling and are not concerned with diffusers of light fittings which are attached to the soffit of, or suspended beneath, a ceiling (see Diagram 26).

Lighting diffusers are translucent or open-structured elements that allow light to pass through. They may be part of a luminaire or used below rooflights or other sources of light.

Diagram 26 **Lighting diffuser in relation to ceiling**

See para 6.13

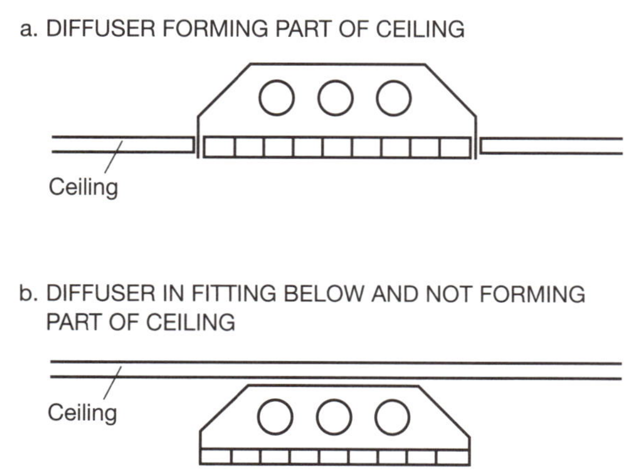

6.14 Thermoplastic lighting diffusers should not be used in fire-protecting or fire-resisting ceilings, unless they have been satisfactorily tested as part of the ceiling system that is to be used to provide the appropriate fire protection.

6.15 Subject to the above paragraphs, ceilings to rooms and circulation spaces (but not protected stairways) may incorporate thermoplastic lighting diffusers if the following provisions are observed:

a. Wall and ceiling surfaces exposed within the space above the suspended ceiling (other than the upper surfaces of the thermoplastic panels) should comply with the general provisions of paragraph 6.1 and Table 10, according to the type of space below the suspended ceiling.

b. If the diffusers are of classification TP(a) (rigid), there are no restrictions on their extent.

c. If the diffusers are of classification TP(b), they should be limited in extent as indicated in Table 11 and Diagram 27.

Suspended or stretched-skin ceilings

6.16 The ceiling of a room may be constructed either as a suspended or as a stretched skin membrane from panels of a thermoplastic material of the TP(a) flexible classification, provided that it is not part of a fire-resisting ceiling. Each panel should not exceed $5m^2$ in area and should be supported on all its sides.

Table 11 Limitations applied to thermoplastic rooflights and lighting diffusers in suspended ceilings and Class 3 plastic rooflights

Minimum classification of lower surface	Use of space below the diffusers or rooflight	Maximum area of each diffuser panel or rooflight [(1)] (m²)	Max total area of diffuser panels and rooflights as percentage of floor area of the space in which the ceiling is located (%)	Minimum separation distance between diffuser panels or rooflights [(1)] (m)
TP(a)	Any except protected stairway	No limit [(2)]	No limit	No limit
Class 3 [(3)] or TP(b)	Rooms	5	50 [(4)(5)]	3 [(5)]
	Circulation spaces except protected stairways	5	15 [(4)]	3

Notes:

1. Smaller panels can be grouped together provided that the overall size of the group and the space between one group and any others satisfies the dimensions shown in Diagram 27.
2. Lighting diffusers of TP(a) flexible rating should be restricted to panels of not more than 5m² each, see paragraph 6.16.
3. There are no limits on Class 3 material in small rooms. See paragraph 6.1, Table 10.
4. The minimum 3m separation specified in Diagram 27 between each 5m² must be maintained. Therefore, in some cases it may not also be possible to use the maximum percentage quoted.
5. Class 3 rooflights to rooms in industrial and other non-residential purpose groups may be spaced 1800mm apart provided the rooflights are evenly distributed and do not exceed 20% of the area of the room.

For industrial buildings and storage buildings such rooflights should be spaced no closer than 3000mm apart.

Diagram 27 Layout restrictions on Class 3 plastic rooflights, TP(b) rooflights and TP(b) lighting diffusers

See Table 11

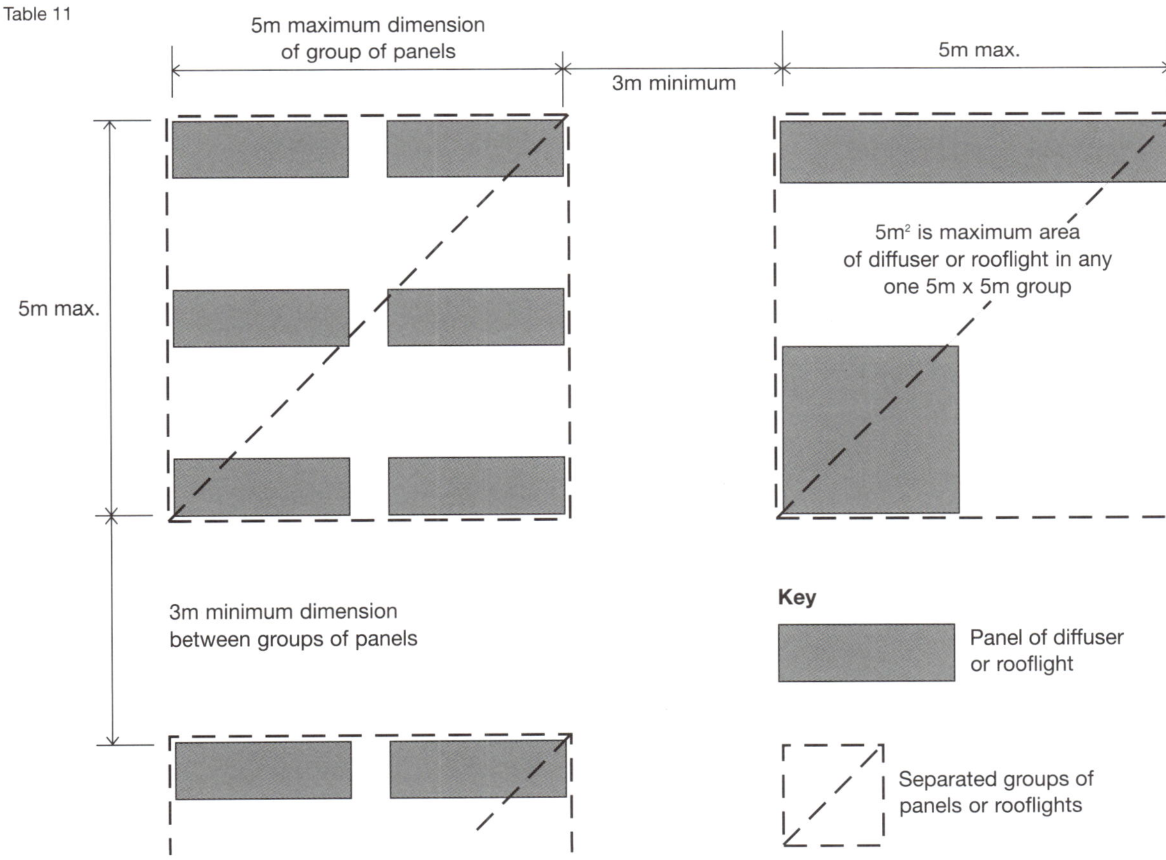

Notes:

a Upper and lower surface of suspended ceiling, between plastic panels, to comply with paragraph 5.1
b No restriction on Class 3 rooflights in small rooms
c See Note 5 to Table 11

B3

The Requirement

This Approved Document deals with the following Requirement from Part B of Schedule 1 to the Building Regulations 2000 (as amended).

Requirement	*Limits on application*
Internal fire spread (structure)	
B3. (1) The building shall be designed and constructed so that, in the event of fire, its stability will be maintained for a reasonable period.	
(2) A wall common to two or more buildings shall be designed and constructed so that it adequately resists the spread of fire between those buildings. For the purposes of this sub-paragraph a house in a terrace and a semi-detached house are each to be treated as a separate building.	
(3) Where reasonably necessary to inhibit the spread of fire within the building, measures shall be taken, to an extent appropriate to the size and intended use of the building, comprising either or both of the following: (a) sub-division of the building with fire-resisting construction; (b) installation of suitable automatic fire suppression systems.	Requirement B3(3) does not apply to material alterations to any prison provided under Section 33 of the Prison Act 1952.
(4) The building shall be designed and constructed so that the unseen spread of fire and smoke within concealed spaces in its structure and fabric is inhibited.	

Insurers' Essential Principles

The relevant essential principles for property protection are:

- Suitable measures will be taken for the prevention of premature structural collapse and excessive deflection;
- The building shall be constructed in such a manner that if a fire starts, the extent of fire and smoke damage will be minimised and confined as close to the source of fire outbreak as is practical/feasible;
- There shall be adequate provision to prevent an arson attack.

Guidance

Performance

In the Secretary of State's view the Requirements of B3 will be met:

a. if the loadbearing elements of structure of the building are capable of withstanding the effects of fire for an appropriate period without loss of stability;

b. if the building is sub-divided by elements of fire-resisting construction into compartments;

c. if any openings in fire-separating elements (see Appendix E) are suitably protected in order to maintain the integrity of the element (i.e. the continuity of the fire separation); and

d. if any hidden voids in the construction are sealed and sub-divided to inhibit the unseen spread of fire and products of combustion, in order to reduce the risk of structural failure and the spread of fire, in so far as they pose a threat to the safety of people in and around the building.

The extent to which any of these measures are necessary is dependent on the use of the building and, in some cases, its size and on the location of the element of construction.

Detailed guidance on compartmentation from a property protection perspective is given in the FPA Design guide: Compartmentation.

Introduction

B3.i Guidance on loadbearing elements of structure is given in Section 7. Section 8 is concerned with the sub-division of a building into compartments and Section 9 makes provisions about concealed spaces (or cavities). Section 10 gives information on the protection of openings and on fire-stopping. Section 11 is concerned with special measures which apply to car parks and shopping complexes. Common to all these sections and to other provisions of Part B, is the property of fire resistance.

Fire resistance

B3.ii The fire resistance of an element of construction is a measure of its ability to withstand the effects of fire in one or more ways, as follows:

a. resistance to collapse, i.e. the ability to maintain loadbearing capacity (which applies to loadbearing elements only);

b. resistance to fire penetration, i.e. an ability to maintain the integrity of the element; and

c. resistance to the transfer of excessive heat, i.e. an ability to provide insulation from high temperatures.

B3.iii "Elements of structure" is the term applied to the main structural loadbearing elements, such as structural frames, floors and loadbearing walls. Compartment walls are treated as elements of structure although they are not necessarily loadbearing. Roofs, unless they serve the function of a floor, are not treated as elements of structure. External walls, such as curtain walls or other forms of cladding which transmit only self weight and wind loads and do not transmit floor load, are not regarded as loadbearing for the purposes of B3.ii(a), although they may need fire resistance to satisfy requirement B4 (see Sections 12 and 13).

Loadbearing elements may or may not have a fire-separating function. Similarly, fire-separating elements may or may not be loadbearing.

Guidance elsewhere in the Approved Document concerning fire resistance

B3.iv There is guidance in Sections 2 to 5 concerning the use of fire-resisting construction to protect means of escape. There is guidance in Section 12 about fire resistance of external walls to restrict the spread of fire between buildings. There is guidance in Section 17 about fire resistance in the construction of firefighting shafts. Appendix A gives information on methods of test and performance for elements of construction. Appendix B gives information on fire doors. Appendix C gives information on methods of measurement. Appendix D gives information on purpose group classification. Appendix E gives definitions.

Section 7: Loadbearing elements of structures

Introduction

7.1 Premature failure of the structure can be prevented by provisions for loadbearing elements of structure to have a minimum standard of fire resistance, in terms of resistance to collapse or failure of loadbearing capacity. The purpose in providing the structure with fire resistance is threefold, namely:

a. to minimise the risk to the occupants, some of whom may have to remain in the building for some time while evacuation proceeds if the building is a large one;

b. to reduce the risk to firefighters, who may be engaged on search or rescue operations; and

c. to reduce the danger to people in the vicinity of the building, who might be hurt by falling debris or as a result of the impact of the collapsing structure on other buildings.

Fire resistance standard

7.2 Elements of structure such as structural frames, beams, columns, loadbearing walls (internal and external), floor structures and gallery structures, should have at least the fire resistance given in Appendix A, Table A1.

Application of the fire resistance standards for loadbearing elements

7.3 The measures set out in Appendix A include provisions to ensure that where one element of structure supports or gives stability to another element of structure, the supporting element has no less fire resistance than the other element (see notes to Table A2). The measures also provide for elements of structure that are common to more than one building or compartment, to be constructed to the standard of the greater of the relevant provisions. Special provisions about fire resistance of elements of structure in single storey buildings are also given and there are concessions in respect of fire resistance of elements of structure in basements where at least one side of the basement is open at ground level.

Exclusions from the provisions for elements of structure

7.4 The following are excluded from the definition of element of structure for the purposes of these provisions:

a. a structure that only supports a roof, unless:

   i. the roof performs the function of a floor, such as for parking vehicles, or as a means of escape (see Sections 2 to 4); or

   ii. the structure is essential for the stability of an external wall which needs to have fire resistance;

   iii. the collapse of the roof makes any compartmentation directly below the roof ineffective;

   This is a potential problem in single-storey buildings where collapse of the roof above the fire can reduce the effectiveness of any compartment or separating wall. More guidance is given in the FPA Design guide: Compartmentation.

b. the lowest floor of the building;

c. a platform floor; and

d. a loading gallery, fly gallery, stage grid, lighting bridge, or any gallery provided for similar purposes or for maintenance and repair (see definition of "Element of structure" in Appendix E).

Additional guidance

7.5 Guidance in other sections of this Approved Document may also apply if a loadbearing wall is:

a. a compartment wall (this includes a wall common to two buildings), (see Section 8);

b. a wall enclosing a place of special fire hazard (see Section 8, paragraph 8.12);

c. protecting a means of escape, (see Sections 2 to 5);

d. an external wall, (see Sections 12 and 13); or

e. enclosing a firefighting shaft, (see Section 17).

7.6 If a floor is also a compartment floor, see Section 8.

Raised storage areas

7.7 Raised free-standing floors (sometimes supported by racking) are frequently erected in single storey industrial and storage buildings. Whether the structure is considered as a gallery or is of sufficient size that it is considered as a floor forming an additional storey, the normal provisions for fire resistance of elements of structure may be onerous if applied to the raised storage area.

7.8 A structure which does not have the appropriate fire resistance given in Appendix A, Table A1 is acceptable provided the following conditions are satisfied:

a. the structure has only one tier and is used for storage purposes **only;**

b. the number of persons likely to be on the floor at any one time is low and does not include members of the public;

c. the floor is not more than 10m in either width or length and does not exceed one half of the floor area of the space in which it is situated;

d. the floor is open above and below to the room or space in which it is situated; and

e. the means of escape from the floor meets the relevant provisions in Sections 3, 4 and 5.

Note 1: Where the lower level is provided with an automatic detection and alarm system meeting the relevant recommendations of BS 5839-1:2002, then the floor size may be increased to not more than 20m in either width or length.

Note 2: The maximum dimensions provided above have been set in order to limit the distance that a firefighter may need to travel over or under the floor to effect a rescue. Where agreed locally it may be possible to vary these dimensions, however, the safety of firefighters who may be on or underneath these floors must be taken into account.

Note 3: Where the building is fitted throughout with an automatic sprinkler system in accordance with paragraph 0.16, there are no limits on the size of the floor.

Note 4: An unprotected raised floor, even when limited to storage only, may collapse in a fire and hinder firefighting operations by blocking access because of fallen goods and debris. Although a size limit is given, if a number of discrete units of unprotected raised floors are used, they are permitted to take up to nearly 50% of the floor area. If the aggregate floor area provided by the raised floors exceeds 500m^2 and the building is not sprinkler protected, then the structure to the raised floor should be protected to at least 30 minutes standard of fire resistance.

Conversion to flats

7.9 Where an existing house or other building is converted into flats, there is a material change of use to which Part B of the regulations applies. Where the existing building has timber floors and these are to be retained, the relevant provisions for fire resistance may be difficult to meet.

7.10 Provided that the means of escape conform to Section 3 and are adequately protected, a 30 minute standard of fire resistance could be accepted for the elements of structure in a building having not more than three storeys.

Where the altered building has four or more storeys the full standard of fire resistance given in Appendix A would normally be necessary.

Section 8: Compartmentation

Introduction

8.1 The spread of fire within a building can be restricted by sub-dividing it into compartments separated from one another by walls and/or floors of fire-resisting construction. The object is twofold:

a. to prevent rapid fire spread which could trap occupants of the building; and

b. to reduce the chance of fires becoming large, on the basis that large fires are more dangerous, not only to occupants and fire and rescue service personnel, but also to people in the vicinity of the building.

c. To contain the fire as far as possible to one compartment for the duration of the fire, thus minimising the damage to the property and as far as is practical minimising business interruption. For the purpose of property protection and business resilience, this may be achieved by a combination of higher fire resistance, smaller compartment areas or requiring the installation of an automatic sprinkler system.

From a property protection perspective, compartmentation is seen as vital to limit the spread of fire in order to:

- *prevent excessive damage caused by fire and smoke and thus limit the financial exposure. Fire and smoke spread outside the room or compartment of origin is to be minimised for the full duration of a fire;*
- *protect valuable materials, equipment and processes. These can represent a significant financial risk both from direct loss and also subsequent business interruption; and*
- *limit the area of damage from a fire in order to minimise business interruption and consequent financial exposure. Many businesses that suffer a fire may not survive because of loss of customers during the time taken to refurbish and replenish stock.*

Compartmentation is complementary to provisions made in Sections 2 to 5 for the protection of escape routes and to provisions made in Sections 12 to 14 against the spread of fire between buildings.

8.2 The appropriate degree of sub-division depends on:

a. the use of and fire load in the building, which affects the potential for fires and the severity of fires, as well as the ease of evacuation;

b. the height to the floor of the top storey in the building, which is an indication of the ease of evacuation and the ability of the fire and rescue service to intervene effectively; and

c. the availability of a sprinkler system which affects the growth rate of the fire and may suppress it altogether.

8.3 Sub-division is achieved using compartment walls and compartment floors. The circumstances in which they are needed are given in paragraphs 8.9 to 8.19.

8.4 Provisions for the construction of compartment walls and compartment floors are given in paragraphs 8.20 onwards. These construction provisions vary according to the function of the wall or floor.

Special forms of compartmentation

8.5 Special forms of compartmentation to which particular construction provisions apply, are:

a. walls common to two or more buildings, see paragraph 8.10;

b. walls dividing buildings into separated parts, see paragraph 8.11; and

c. construction enclosing places of special fire hazard, see paragraph 8.12.

Junctions

8.6 For compartmentation to be effective, there should be continuity at the junctions of the fire-resisting elements enclosing a compartment and any openings from one compartment to another should not present a weakness.

Fire resistance can be significantly reduced by any weakness at junctions of fire-resisting compartments and openings.

Protected shafts

8.7 Spaces that connect compartments, such as stairways and service shafts, need to be protected to restrict fire spread between the compartments and they are termed protected shafts. Any walls or floors bounding a protected shaft are considered to be compartment walls or floors for the purpose of this Approved Document.

Buildings containing one or more atria

8.8 Detailed advice on all issues relating to the incorporation of atria in buildings is given in BS 5588-7:1997. However, it should be noted that for the purposes of Approved Document B, the standard is relevant **only where the atrium breaches any compartmentation**.

Such breaching of compartmentation potentially increases the amount of damage due to smoke and fire. Where floors are breached in this way, it is essential to provide alternative ways of containing fire and smoke to the compartment of origin, for example by using a combination of sprinklers and smoke extraction or enclosing the perimeter of the atria in fire-resisting glazing.

Provision of compartmentation

General

8.9 Compartment walls and compartment floors should be provided in the circumstances described below, with the proviso that the lowest floor in a building does not need to be constructed as a compartment floor. Paragraphs 8.10 to 8.19 give guidance on the provision of compartmentation in different building types. Information on the construction of compartment walls and compartment floors in different circumstances is given in paragraphs 8.20 to 8.31. Provisions for the protection of openings in compartment walls and compartment floors are given in paragraphs 8.32 to 8.34.

All purpose groups

8.10 A wall common to two or more buildings should be constructed as a compartment wall.

8.11 Parts of a building that are occupied mainly for different purposes should be separated from one another by compartment walls and/or compartment floors. This does not apply where one of the different purposes is ancillary to the other. Refer to Appendix D for guidance on whether a function should be regarded as ancillary or not.

Places of special fire hazard

8.12 Every place of special fire hazard (see Appendix E) should be enclosed with fire-resisting construction; see Table A1, Item 13.

Note: Any such walls and floors are not compartment walls and compartment floors.

Flats

8.13 In buildings containing flats, the following should be constructed as compartment walls or compartment floors:

a. every floor (unless it is within a flat, i.e. between one storey and another within one individual dwelling); and

b. every wall separating a flat from any other part of the building; and

   Note: Any other part of the building does not include an external balcony/deck access.

c. every wall enclosing a refuse storage chamber.

8.14 Blocks of flats with a floor more than 30m above ground level should be fitted with a sprinkler system in accordance paragraph 0.16.

Note: Sprinklers need only be provided within the individual flats, they are not required in the common areas such as stairs, corridors or landings. For the purposes of this paragraph the limit on the scope of BS 9251:2005 to buildings below 20m in height can be ignored.

Institutional buildings including health care

8.15 All floors should be constructed as compartment floors.

8.16 Paragraphs 3.41 to 3.52 give guidance eon the provisions for compartment walls in care homes utilising progressive horizontal evacuation.

Other residential buildings

8.17 All floors should be constructed as compartment floors.

Non-residential buildings

8.18 The following walls and floors should be constructed as compartment walls and compartment floors in buildings of a non-residential purpose group (i.e. Office, Shop and Commercial, Assembly and Recreation, Industrial, Storage or Other non-residential):

Ideally, all floors should be constructed as compartment floors. Where a floor contains an unprotected opening, or accommodation stair, allowable maximum areas for each floor in the compartment may need to be reduced (see Diagram to Table 12a).

a. every wall needed to sub-divide the building to observe the size limits on compartments given in Table 12/Table 12a (see Diagram 28a);

b. every floor, if the building or separated part (see paragraph 8.22) of the building, has a storey with a floor at a height of more than 30m above ground level (see Diagram 28b); (20m is preferred for property protection, see Diagram 28.)

c. the floor of the ground storey if the building has one or more basements (see Diagram 28c), with the exception of small premises (see paragraph 3.1);

d. the floor of every basement storey (except the lowest floor) if the building, or separated part (see paragraph 8.19), has a basement at a depth of more than 10m below ground level (see Diagram 28d);

e. if the building forms part of a shopping complex, every wall and floor described in Section 5 of BS 5588-0:1991 *Fire precautions in the design, construction and use of buildings, Code of practice for shopping complexes* as needing to be constructed to the standard for a compartment wall or compartment floor; and

f. if the building comprises Shop and Commercial, Industrial or Storage premises, every wall or floor provided to divide a building into separate occupancies, (i.e. spaces used by different organisations whether they fall within the same Purpose Group or not).

Note: See also the provision in paragraph 5.58 for store rooms in shops to be separated from retail areas by fire-resisting construction to the standard given in Table A1.

8.19 In a two storey building in the Shop and Commercial or Industrial Purpose Groups, where the use of the upper storey is ancillary to the use of the ground storey, the ground storey may be treated as a single storey building for fire compartmentation purposes, provided that:

a. the area of the upper storey does not exceed 20% of the area of the ground storey, or 500m^2, whichever is less;

b. the upper storey is compartmented from the lower one; and

c. there is a means of escape from the upper storey that is independent of the routes from the lower storey.

Diagram 28 **Compartment floors: illustration of guidance in paragraph 8.18**

A. EXAMPLE OF COMPARTMENTATION IN AN UNSPRINKLERED SHOP see paragraph 8.18(a)

None of the floors in this case would need to be compartment floors, but the two storeys exceeding 2000m^2 would need to be divided into compartments not more than 2000m^2 by compartment walls.

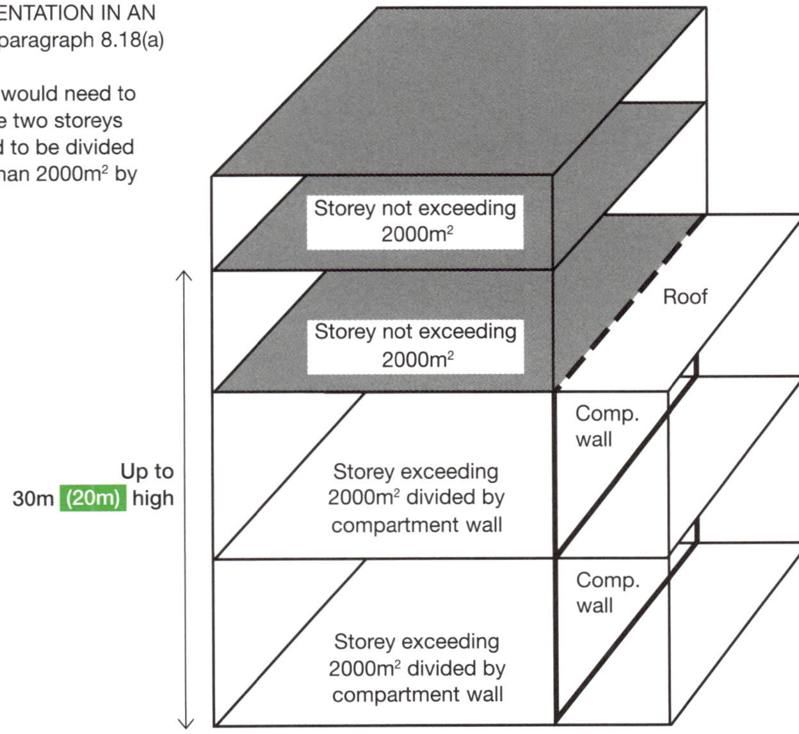

The compartment walls in example (a) do not need to be in one vertical plane

B. COMPARTMENTATION IN TALL BUILDINGS see paragraph 8.18(b)

In a building over 30m (20m) in height all storeys should be separated by compartment floors. For advice on the special conditions in atrium buildings see BS 5588-7

C. SHALLOW BASEMENT see paragraph 8.18(c)

Only the floor of the ground storey need be a compartment floor if the lower basement is at a depth of not more than 10m

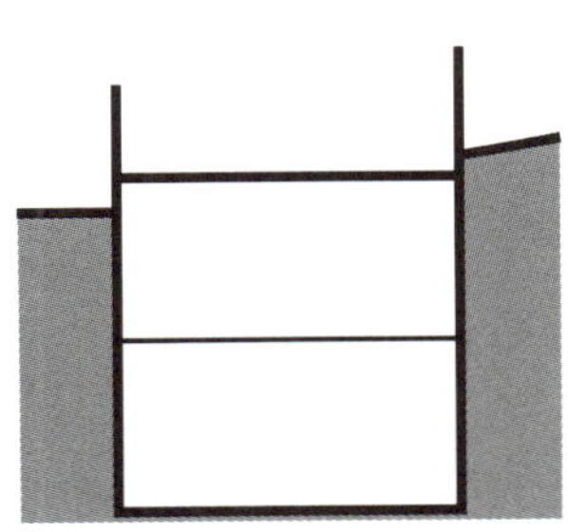

D. DEEP BASEMENTS see paragraph 8.18(d)

All basement storeys to be separated by compartment floors if any storey is at a depth of more than 10m

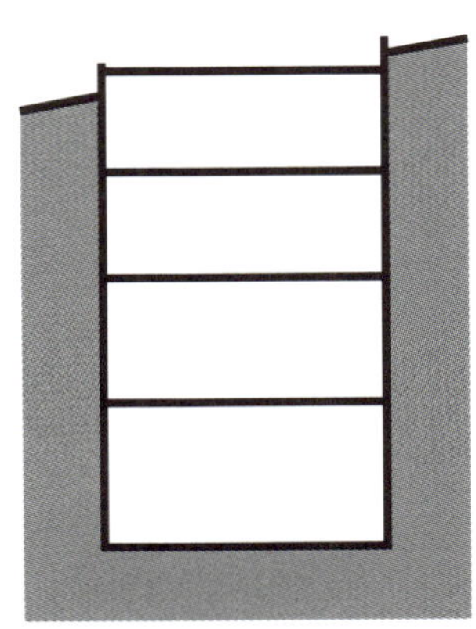

Table 12 **Maximum dimensions of building or compartment (non-residential buildings)**

Purpose Group of building or part	Height of floor of top storey above ground level (m)	Floor area of any one storey in the building or any one storey in a compartment (m^2): In multi-storey buildings	Floor area of any one storey in the building or any one storey in a compartment (m^2): In single-storey buildings
Office	No limit	No limit	No limit
Assembly and recreation Shop and commercial:			
a. Shops – not sprinklered	No limit	2000	2000
Shops – sprinklered (1)	No limit	4000	No limit
b. Elsewhere – not sprinklered	No limit	2000	No limit
Elsewhere – sprinklered (1)	No limit	4000	No limit
Industrial (2)			
Not sprinklered	Not more than 18 More than 18	7000 2000 (3)	No limit N/A
Sprinklered (1)	Not more than 18 More than 18	14,000 4000 (3)	No limit N/A

	Height of floor of top storey above ground level (m)	maximum compartment volume m^3 (multi-storey buildings)	maximum floor area (m^2) (single-storey buildings)	maximum height (m) (4) (single-storey buildings)
Storage (2) and other non-residential:				
a. Car park for light vehicles	No limit	No limit	No limit	No limit
b. Any other building or part:				
Not sprinklered	Not more than 18 More than 18	20,000 4000(3)	20,000 N/A	18 N/A
Sprinklered (1)	Not more than 18 More than 18	40,000 8000(3)	No limit	No limit

Notes:

1. 'Sprinklered' means that the building is fitted throughout with an automatic sprinkler in accordance with paragraph 0.16.
2. There may be additional limitations on floor area and/or sprinkler provisions in certain industrial and storage uses under other legislation, for example in respect of storage of LPG and certain chemicals, or given by insurers (see Table 12a/Table 12b).
3. This reduced limit applies only to storeys that are more than 18m above ground level. Below this height the higher limit applies.
4. Compartment height is measured from finished floor level to underside of roof or ceiling.

Table 12a Guidance on maximum dimensions of building or compartment (non-residential buildings) for property protection

These values are for guidance only, as other factors such as value of building contents may dictate further restrictions on floor areas, particularly when building is not sprinkler protected

Purpose group of building or part	Height of floor of top storey above ground level (m)	Floor area of any compartment[5] [6] (m²): In multi-storey buildings	Floor area of any compartment[5] [6] (m²): In single-storey buildings
Office – not sprinklered	20m	4000	No limit
Office – sprinklered	No limit	No limit	No limit
Assembly and recreation Shop and commercial:			
a. Shops – not sprinklered	No limit	2000	2000
Shops – sprinklered [1]	No limit	4000	8000
b. Elsewhere – not sprinklered	No limit	2000	4000
Elsewhere – sprinklered [1]	No limit	4000	No limit
Industrial [2]			
Not sprinklered	Not more than 18 More than 18	4000 2000 [3]	7000 4000
Sprinklered [1]	Not more than 18 More than 18	7000 4000 [3]	14000 7000

	Height of floor of top storey above ground level (m)	maximum compartment volume (m³) – multi-storey buildings	maximum floor area (m²) – single-storey buildings	maximum height (m) [4] – single-storey buildings
Storage [2] and other non-residential:				
a. Car park for light vehicles	No limit	No limit	No limit	No limit
b. Any other building or part:				
Not sprinklered	Not more than 18 More than 18	See Table 12b		
Sprinklered [1]	Not more than 18 More than 18			

Notes:

1. 'Sprinklered' means that the building is fitted throughout with an automatic sprinkler in accordance with paragraph 0.16
2. There may be additional limitations on floor area and/or sprinkler provisions in certain industrial and storage uses under other legislation, for example in respect of storage of LPG and certain chemicals.
3. This reduced limit applies only to storeys that are more than 18m above ground level. Below this height the higher limit applies.
4. Compartment height is measured from finished floor level to underside of roof (eaves) or ceiling.
5. In determining maximum acceptable floor area and any height limitation, insurers will consider use of building, type and nature of contents and total sum insured (TSI). TSI would include not only property damage (PD) values but also business interruption (BI). Building designers are advised to speak to the insurer at the early stage of the design to obtain guidance.
6. With regard to multi-storey buildings, compartment floor area is defined for property protection purposes as the maximum total allowable compartment floor area between floors constructed as compartment floors (see Diagram to Table 12a).

Table 12b Maximum floor areas for specific applications in warehouses and storage buildings – subject to insurers' individual risk assessment

Application (Roof height less than 10m [1])	Recommended floor areas (m^2) [2]		
	High risk	Medium risk	Low risk
Warehouse with high rack storage	2000	4000	8000
Storage building connected to factory/ production area without high rack storage	4000	8000	10000
Areas containing hazardous processes, equipment or other critical areas, including quarantine areas	250	500	

Notes:

1. Warehouses with a ceiling height greater than 10m present a particular risk. Proposals for such buildings should be referred to the insurer for specific guidance on requirements.
2. Recommended floor areas may be increased if a suitable automatic extinguishing system is provided. Whether or not an increase is allowable, and the amount of increase, will be subject to the insurers' individual risk assessment.
3. For definition of high risk, medium risk and low risk, see the *FPA Design guide: Warehouses.*

Diagram to Table 12a

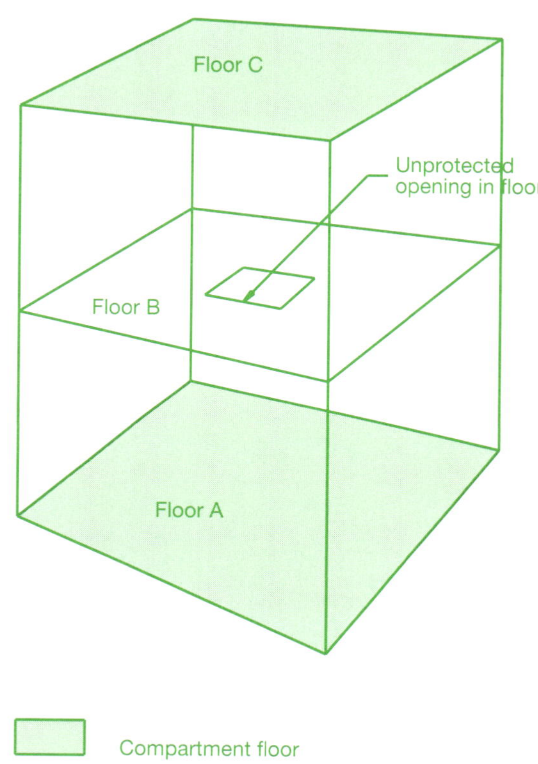

Floors A and C are constructed as compartment floors. Floor B contains an unprotected opening for an accommodation staircase. Thus the total combined area of floors A and B should not exceed the appropriate value given in Table 12a.

Factors that insurers will take into account when recommending limits on floor area include the following:

- ***proposed floor area and cubic capacity of the building;***
- ***occupancy;***
- ***location;***
- ***fire load and nature of the goods stored;***
- ***nature and extent of any packaging used e.g. warehouses;***
- ***construction;***
- ***general fire safety management and control features;***
- ***value at risk, but more particularly, the overall estimated maximum loss (EML) in respect of buildings, contents and business interruption; and***
- ***reinsurance aspects.***

Construction of compartment walls and compartment floors

General

8.20 Every compartment wall and compartment floor should:

a. form a complete barrier to fire between the compartments they separate; and

b. have the appropriate fire resistance as indicated in Appendix A, Tables A1 and A2.

Note 1: Timber beams, joists, purlins and rafters may be built into or carried through a masonry or concrete compartment wall if the openings for them are kept as small as practicable and then fire-stopped. If trussed rafters bridge the wall, they should be designed so that failure of any part of the truss due to a fire in one compartment will not cause failure of any part of the truss in another compartment.

Note 2: Where services are incorporated within the construction that could provide a potential source of ignition, care should be taken to ensure the risk of fire developing and spreading prematurely into adjacent compartments is controlled.

Compartment walls between buildings

8.21 Compartment walls that are common to two or more buildings should run the full height of the building in a continuous vertical plane. Thus adjoining buildings should only be separated by walls, not floors.

Separated parts of buildings

8.22 Compartment walls used to form a separated part of a building (so that the separated parts can be assessed independently for the purpose of determining the appropriate standard of fire resistance) should run the full height of the building in a continuous vertical plane. The two separated parts can have different standards of fire resistance.

Other compartment walls

8.23 Compartment walls not described in paragraphs 8.21 and 8.22 should run the full height of the storey in which they are situated.

8.24 Compartment walls in a top storey beneath a roof should be continued through the roof space (see definition of compartment in Appendix E).

Junction of compartment wall or compartment floor with other walls

8.25 Where a compartment wall or compartment floor meets another compartment wall or an external wall, the junction should maintain the fire resistance of the compartmentation. Fire-stopping should meet the provisions of paragraphs 10.17 to 10.19.

8.26 At the junction of a compartment floor with an external wall that has no fire resistance (such as a curtain wall) the external wall should be restrained at floor level to reduce the movement of the wall away from the floor when exposed to fire.

8.27 Compartment walls should be able to accommodate the predicted deflection of the floor above by either:

a. having a suitable head detail between the wall and the floor, that can deform but maintain integrity when exposed to a fire; or

b. the wall may be designed to resist the additional vertical load from the floor above as it sags under fire conditions and thus maintain integrity.

Note: Where compartment walls are located within the middle half of a floor between vertical supports, the predicted deflection may be assumed to be 40mm unless a smaller value can be justified by assessment. Outside this area the limit can be reduced linearly to zero at the supports. For steel beams that do not have the required fire resistance, reference should be made to SCI Publication 288 *Fire safe design: A new approach to multi-storey steel-framed buildings* (Second Edition) 2000 (ISBN: 1 85942 169 5).

Diagram 29 **Compartment walls and compartment floors with reference to relevant paragraphs in Section 8**

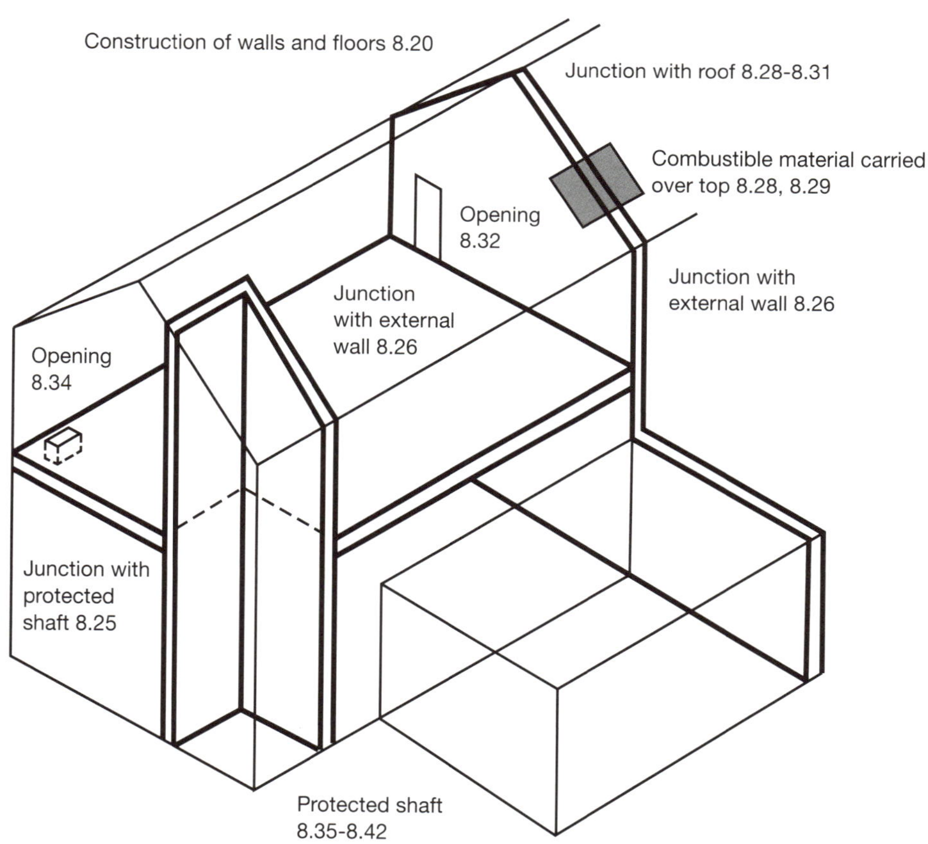

Junction of compartment wall with roof

8.28 A compartment wall should be taken up to meet the underside of the roof covering or deck, with fire-stopping where necessary at the wall/roof junction to maintain the continuity of fire resistance. The compartment wall should also be continued across any eaves cavity (see paragraph 8.20a).

8.29 If a fire penetrates a roof near a compartment wall there is a risk that it will spread over the roof to the adjoining compartment. To reduce this risk and subject to paragraph 8.30, a zone of the roof 1500mm wide on either side of the wall should have a covering of designation AA, AB or AC (see Appendix A, paragraph 6) on a substrate or deck of a material of limited combustibility, as set out in Diagram 30a.

Note: Thermoplastic rooflights which, by virtue of paragraph 14.7, are regarded as having an AA *(National class) designation or* B_{ROOF}*(t4) (European class) classification* are **not** suitable for use in the zone described above (or within the protected zone).

8.30 In buildings not more than 15m high, of the purpose groups listed below, combustible boarding used as a substrate to the roof covering, wood wool slabs, or timber tiling battens, may be carried over the compartment wall provided that they are fully bedded in mortar or other suitable material over the width of the wall (see Diagram 30b). This applies to, buildings or compartments in Residential use (other than Institutional), Office buildings, Assembly and Recreation buildings.

Note: Double-skinned insulated roof sheeting, with a thermoplastic core, should incorporate a band of material of limited combustibility at least 300mm wide centred over the wall.

8.31 As an alternative to paragraphs 8.29 or 8.30 the compartment wall may be extended up through the roof for a height of at least 375mm above the top surface of the adjoining roof covering. Where there is a height difference of at least 375mm between two roofs or where the roof coverings on either side of the wall are AA, AB or AC this height may be reduced to 200mm (see Diagram 30c).

Openings in compartmentation

Openings in compartment walls separating buildings or occupancies

8.32 Any openings in a compartment wall which is common to two or more buildings, or between different occupancies in the same building, should be limited to those for:

a. a door which is needed to provide a means of escape in case of fire and which has the same fire resistance as that required for the wall (see Appendix B, Table B1) and is fitted in accordance with the provisions of Appendix B; and

b. the passage of a pipe which meets the provisions in Section 10.

Doors

8.33 Information on fire doors may be found in Appendix B.

Openings in other compartment walls or in compartment floors

8.34 Openings in compartment walls (other than those described in paragraph 8.32) or compartment floors should be limited to those for:

a. doors which have the appropriate fire resistance given in Appendix B, Table B1 and are fitted in accordance with the provisions of Appendix B;

b. the passage of pipes, ventilation ducts, service cables, chimneys, appliance ventilation ducts or ducts encasing one or more flue pipes, which meet the provisions in Section 9;

c. refuse chutes of non-combustible construction;

d. atria designed in accordance with BS 5588-7:1997; and

e. protected shafts which meet the relevant provisions below.

Protected shafts

8.35 Any stairway or other shaft passing directly from one compartment to another should be enclosed in a protected shaft so as to delay or prevent the spread of fire between compartments.

There are additional provisions in Sections 2 to 5 for protected shafts that are protected stairways and in Section 17 if the stairway also serves as a firefighting stair.

Diagram 30 **Junction of compartment wall with roof**

See paras 8.28–8.31

a. ANY BUILDING OR COMPARTMENT

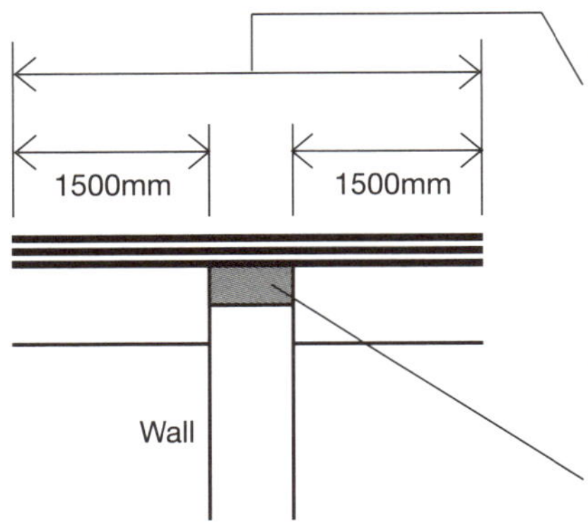

Roof covering over this distance to be designated AA, AB or AC on deck of material of limited combustibility. Roof covering and deck could be composite structure, e.g. profiled steel cladding.

Double-skinned insulated roof sheeting should incorporate a band of material of limited combustibility at least 300mm wide centred over the wall.

If roof support members pass through the wall, fire protection to these members for a distance of 1500mm on either side of the wall may be needed to delay distortion at the junction (see note to paragraph 8.20).

Resilient fire-stopping to be carried up to underside of roof covering. e.g. roof tiles.

b. RESIDENTIAL (NOT INSTITUTIONAL), OFFICE OR ASSEMBLY USE AND NOT MORE THAN 15M HIGH

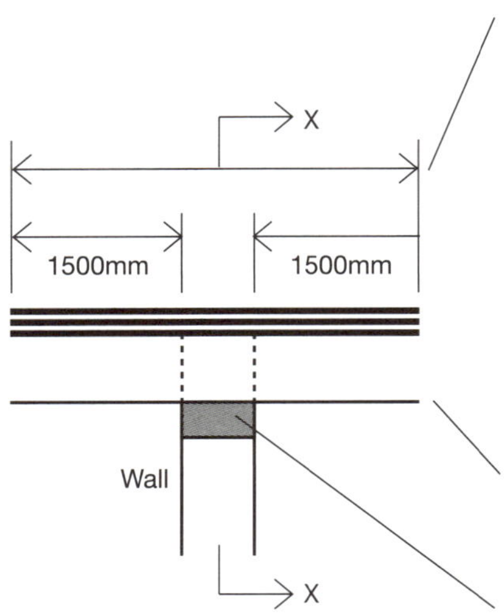

Roof covering to be designated AA, AB or AC for at least this distance.

Boarding (used as a substrate), wood wool slabs or timber tiling battens may be carried over the wall provided that they are fully bedded in mortar (or other no less suitable material) where over the wall.

Thermoplastic insulation materials should not be carried over the wall.

Double-skinned insulated roof sheeting with a thermoplastic core should incorporate a band of material of limited combustibility at least 300mm wide centred over the wall.

Sarking felt may also be carried over the wall.

If roof support members pass through the wall, fire protection to these members for a distance of 1500mm on either side of the wall may be needed to delay distortion at the junction (see note to paragraph 8.20).

Fire-stopping to be carried up to underside of roof covering, boarding or slab.

Section X–X

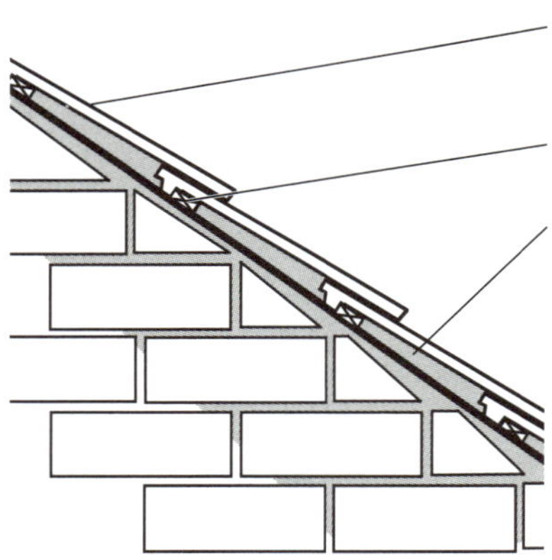

Roof covering to be designated AA, AB or AC for at least 1500mm either side of wall.

Roofing battens and sarking felt may be carried over the wall.

Fire-stopping to be carried up to underside of roof covering. Above and below sarking felt.

Notes

1 Fire-stopping should be carried over the full thickness of the wall.
2 Fire-stopping should be extended into any eaves.
3 The compartment wall need not necessarily be constructed of masonry.

c. ANY BUILDING OR COMPARTMENT

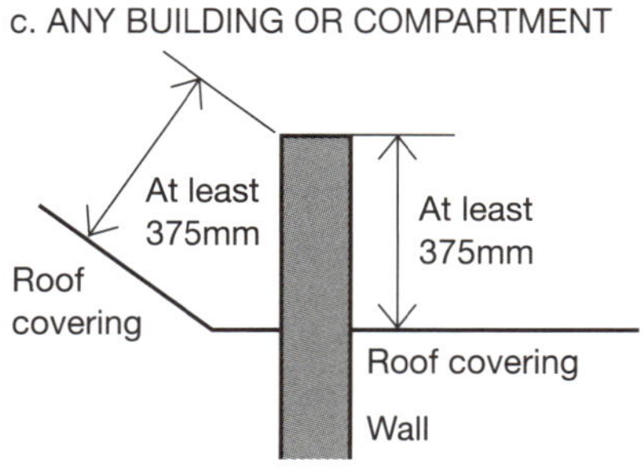

The wall should be extended up through the roof for a height of at least 375mm above the top surface of the adjoining roof covering.

Where there is a height difference of at least 375 mm between two roofs or where the roof coverings on either side of the wall are AA, AB or AC the height of the upstand/parapet wall above the highest roof may be reduced to 200mm.

Uses for protected shafts

8.36 The uses of protected shafts should be restricted to stairs, lifts, escalators, chutes, ducts and pipes. Sanitary accommodation and washrooms may be included in protected shafts.

Construction of protected shafts

8.37 The construction enclosing a protected shaft (see Diagram 31) should:

a. form a complete barrier to fire between the different compartments which the shaft connects;

b. have the appropriate fire resistance given in Appendix A, Table A1, except for uninsulated glazed screens which meet the provisions of paragraph 8.38; and

c. satisfy the provisions about their ventilation and the treatment of openings in paragraphs 8.41 and 8.42.

Uninsulated glazed screens to protected shafts

8.38 If the conditions given below and described in Diagram 32 are satisfied, an uninsulated glazed screen may be incorporated in the enclosure to a protected shaft between a stair and a lobby or corridor which is entered from the stair. The conditions to be satisfied are:

a. the standard of fire resistance for the stair enclosure is not more than 60 minutes; and

b. the glazed screen:

  i. has at least 30 minutes fire resistance in terms of integrity; and

  ii. meets the guidance in Appendix A, Table A4, on the limits on areas of uninsulated glazing; and

c. the lobby or corridor is enclosed to at least a 30 minute standard.

8.39 Where the measures in Diagram 32 to protect the lobby or corridor are not provided, the enclosing walls should comply with Appendix A, Table A1 (item 8c) and the doors with the guidance in Appendix A, Table A4.

Diagram 31 **Protected shafts**

See paras 8.35–8.37

Protected shafts provide for the movement of people (e.g. stairs, lifts), or for passage of goods, air or services such as pipes or cables between different compartments. The elements enclosing the shaft (unless formed by adjacent external walls) are compartment walls and floors. The diagram shows three common examples which illustrate the principles.

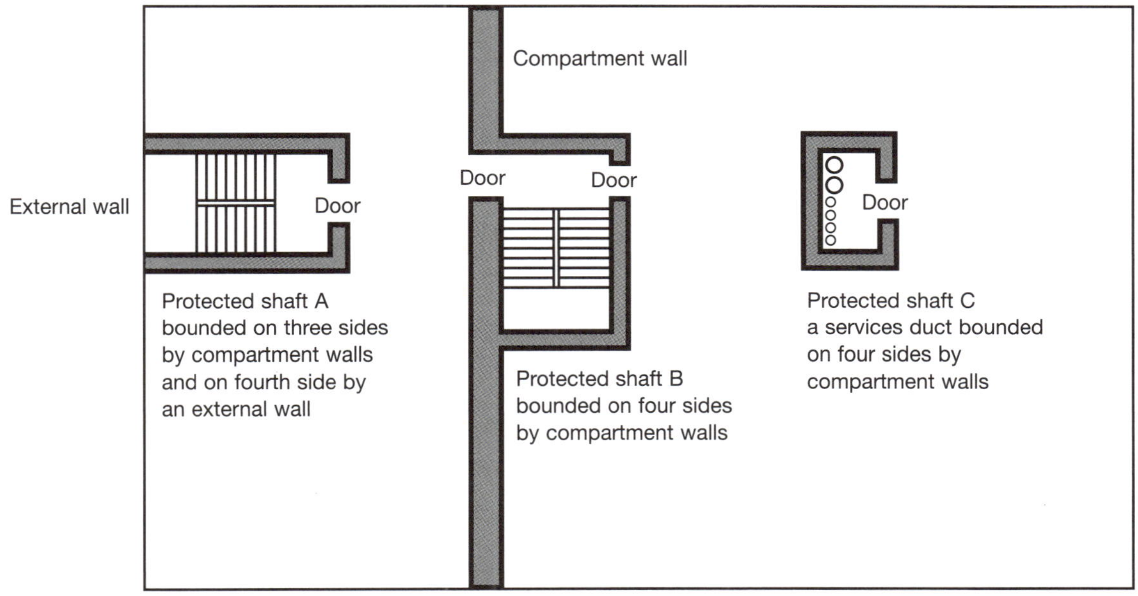

The shaft structure (including any openings) should meet the relevant provisions for: compartment walls (see paragraphs 8.20 to 8.34), external walls (see Sections 12 and 13 and Diagram 24).

Diagram 32 Uninsulated glazed screen separating protected shaft from lobby or corridor

See paras 8.38–8.39

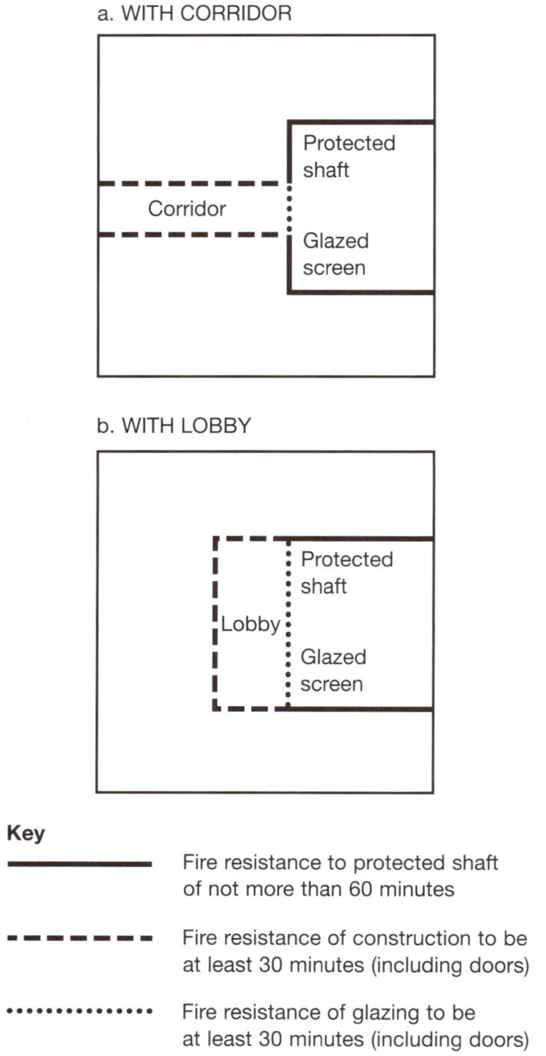

Pipes for oil or gas and ventilation ducts in protected shafts

8.40 If a protected shaft contains a stair and/or a lift, it should not also contain a pipe conveying oil (other than in the mechanism of a hydraulic lift) or contain a ventilating duct (other than a duct provided for the purposes of pressurizing the stairway to keep it smoke free; or a duct provided solely for ventilating the stairway).

Any pipe carrying natural gas or LPG in such a shaft should be of screwed steel or of all welded steel construction, installed in accordance with the *Pipelines Safety Regulations 1996,* SI 1996 No 825 and the *Gas Safety (Installation and use) Regulations 1998,* SI 1998 No 2451.

Note: A pipe is not considered to be contained within a protected shaft if the pipe is completely separated from that protected shaft by fire-resisting construction.

Any pipe containing volatile fluids would need to be contained within a fire-resisting construction of not less than 60 minutes in terms of both integrity and insulation when tested to an appropriate standard.

Ventilation of protected shafts conveying gas

8.41 A protected shaft conveying piped flammable gas should be adequately ventilated direct to the outside air by ventilation openings at high and low level in the shaft.

Any extension of the storey floor into the shaft should not compromise the free movement of air over the entire length of the shaft. Guidance on such shafts, including sizing of the ventilation openings, is given in BS 8313:1997.

Openings into protected shafts

8.42 Generally an external wall of a protected shaft does not need to have fire resistance.

However, there are some provisions for fire resistance of external walls of firefighting shafts in BS 5588-5:2004, which is the relevant guidance called up by paragraph 17.11 to 17.13 and of external walls to protected stairways (which may also be protected shafts) in paragraph 5.24.

Openings in other parts of the enclosure to a protected shaft should be limited as follows:

a. Where part of the enclosure to a protected shaft is a wall common to two or more buildings, only the following openings should be made in that wall:

   i. a door which is needed to provide a means of escape in case of fire; and which has the same fire resistance as that required for the wall (see Appendix B, Table B1); and is fitted in accordance with the provisions of Appendix B; and/or

   ii. the passage of a pipe which meets the provisions in Section 10.

b. Other parts of the enclosure (other than an external wall) should only have openings for:

   i. doors which have the appropriate fire resistance given in Appendix B, Table B1 and are fitted in accordance with the provisions of Appendix B;

   ii. the passage of pipes which meet the provisions in Section 10;

   iii. inlets to, outlets from and openings for a ventilation duct, (if the shaft contains or serves as a ventilating duct) which meet the provisions in Section 10; and/or

   iv. the passage of lift cables into a lift machine room (if the shaft contains a lift). If the machine room is at the bottom of the shaft, the openings should be as small as practicable.

Section 9: Concealed spaces (cavities)

Introduction

9.1 Concealed spaces or cavities in the construction of a building provide a ready route for smoke and flame spread. This is particularly so in the case of voids in, above and below the construction of a building, e.g. walls, floors, ceilings and roofs. As any spread is concealed, it presents a greater danger than would a more obvious weakness in the fabric of the building.

A particular concern is when services pass through a cavity barrier. Services are likely to be added or replaced during the life of the building. The specification of the cavity barrier system should allow for the replacement of services and additional services to be added. It should also facilitate the maintenance of the specified fire resistance. The specifier should ensure that the cavity barrier is suitable for the service and that the fire resistance of the barrier will not be reduced.

Provision of cavity barriers

9.2 Provisions for cavity barriers are given below for specified locations. The provisions necessary to restrict the spread of smoke and flames through cavities are broadly for the purpose of sub-dividing:

a. cavities, which could otherwise form a pathway around a fire-separating element and closing the edges of cavities; therefore reducing the potential for unseen fire spread; and

Note: These should not be confused with fire-stopping details, see Section 10 and Diagram 33 (see also paragraphs 9.3 to 9.7).

b. extensive cavities (see paragraphs 9.8 to 9.12).

Consideration should also be given to the construction and fixing of cavity barriers provided for these purposes and the extent to which openings in them should be protected. For guidance on these issues, see paragraphs 9.13 to 9.16 respectively.

Diagram 33 **Provisions for cavity barriers**

See para 9.2

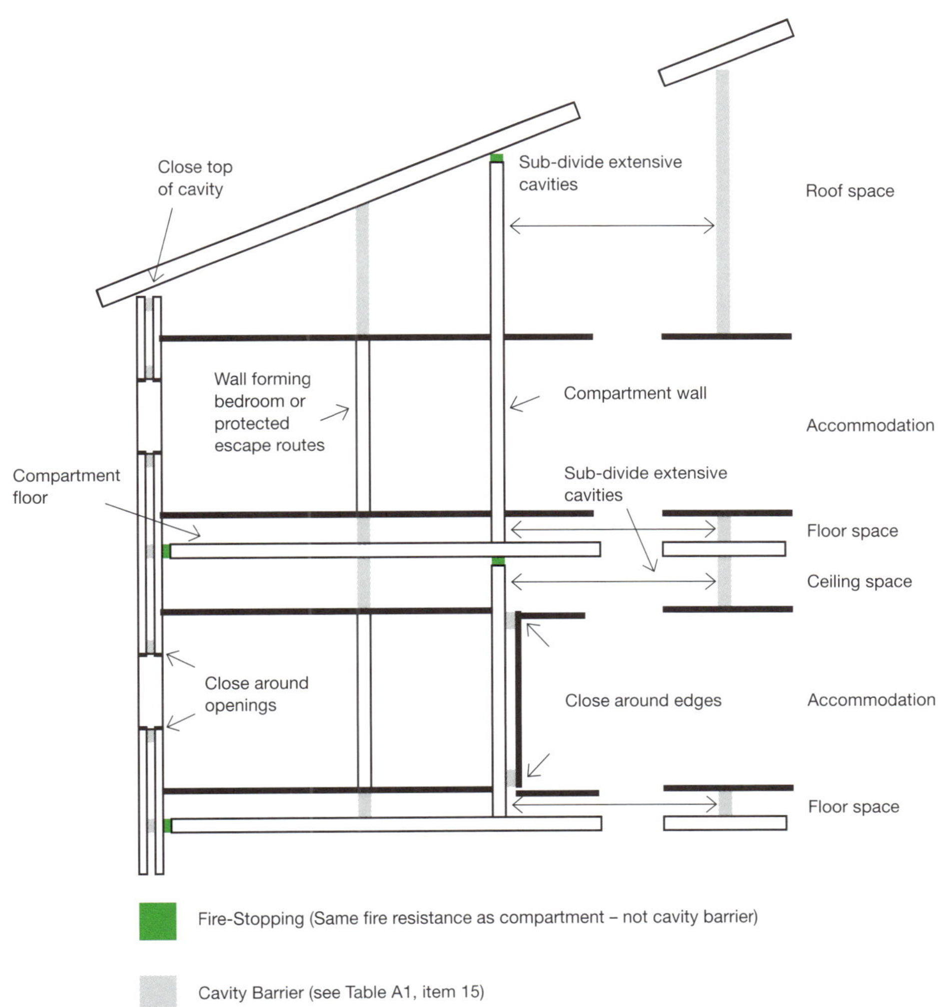

Pathways around fire-separating elements

Junctions and cavity closures

9.3 Cavity barriers should be provided to close the edges of cavities, including around openings.

Cavity barriers should also be provided:

a. at the junction between an external cavity wall (except where the cavity wall complies with Diagram 34) and every compartment floor and compartment wall; and

b. at the junction between an internal cavity wall (except where the cavity wall complies with Diagram 34) and every compartment floor, compartment wall, or other wall or door assembly which forms a fire-resisting barrier.

It is important to continue any compartment wall up through a ceiling or roof cavity to maintain the standard of fire resistance – therefore compartment walls should be carried up full storey height to a compartment floor or to the roof as appropriate, see paragraphs 8.21 to 8.24. It is therefore not appropriate to complete a line of compartmentation by fitting cavity barriers above them.

Protected escape routes

9.4 For a protected escape route, a cavity that exists **above or below** any fire-resisting construction because the construction is not carried to full storey height or, in the case of a top storey, to the underside of the roof covering, should either be:

a. fitted with cavity barriers on the line of the enclosure(s) to the protected escape route; or

b. for cavities above the fire-resisting construction, enclosed on the lower side by a fire-resisting ceiling which extends throughout the building, compartment or separated part (see Diagram 35).

Double-skinned corrugated or profiled roof sheeting

9.5 Cavity Barriers need not be provided between double-skinned corrugated or profiled insulated roof sheeting, if the sheeting is a material of limited combustibility and both surfaces of the insulating layer have a surface spread of flame of at least Class 0 or 1 (National class) or Class C-s3, d2 or better (European class) (see Appendix A) and make contact with the inner and outer skins of cladding (see Diagram 36).

Note: See also paragraph 8.29 Note 2 regarding the junction of a compartment wall with a roof.

Note: When a classification includes "s3, d2", this means that there is no limit set for smoke production and/or flaming droplets/particles.

Cavities affecting alternative escape routes

9.6 Cavitiy barriers may be needed where corridors are be sub-divided to prevent alternative escape routes being simultaneously affected by fire and/or smoke (see paragraph 3.26 and Diagram 16).

Separation of bedrooms

9.7 In Institutional and Other Residential buildings, a cavity that exists above or below partitions between bedrooms because the enclosures are not carried to full storey height, or, (in the case of the top storey) to the underside of the roof covering, should either be:

a. fitted with cavity barriers on the line of the partitions; or

b. for cavities above the partitions, enclosed on the lower side by a fire-resisting ceiling which extends throughout the building, compartment or separated part.

Extensive Cavities

9.8 Cavity barriers should be used to sub-divide any cavity, including any roof space, so that the distance between cavity barriers does not exceed the dimensions given in Table 13.

Maximum dimensions of concealed spaces

9.9 Table 13 sets out maximum dimensions for undivided concealed spaces. With the exceptions given in paragraphs 9.10 to 9.12, extensive concealed spaces should be sub-divided to comply with the dimensions in Table 13.

9.10 The provisions in Table 13 do not apply to any cavity described below:

a. in a wall which should be fire-resisting only because it is loadbearing;

b. in a masonry or concrete external cavity wall shown in Diagram 34;

c. in any floor or roof cavity above a fire-resisting ceiling, as shown in Diagram 35 and which extends throughout the building or compartment subject to a 30m limit on the extent of the cavity; or

d. formed behind the external skin of an external cladding system with a masonry or concrete inner leaf at least 75mm thick, or by overcladding an existing masonry (or concrete) external wall, or an existing concrete roof, provided that the cavity does not contain combustible insulation and the building is not put to a residential or institutional use; or

Table 13 Maximum dimensions of cavities in non-domestic buildings (Purpose Groups 2–7)

Location of cavity	Class of surface/product exposed in cavity (excluding the surface of any pipe, cable or conduit, or any insulation to any pipe)		Maximum dimensions in any direction (m)
	National class	**European class**	
Between roof and a ceiling	Any	Any	20
Any other cavity	Class 0 or Class 1	Class A1 or Class A2-s3, d2 or Class B-s3, d2 or Class C-s3, d2	20
	Not Class 0 or Class 1	Not any of the above classes	10
Cavities deeper than 600mm or above areas of special risk	Class 0	Class A1 or Class A2-s3, d2 or Class B-s3, d2	10

Notes:
1 Exceptions to these provisions are given in paragraphs 9.10 to 9.12.
2 The national classifications do not automatically equate with the equivalent classifications in the European column, therefore, products cannot typically assume a European class unless they have been tested accordingly.
3 When a classification includes "s3, d2", this means that there is no limit set for smoke production and/or flaming droplets/particles.

e. between double-skinned corrugated or profiled insulated roof sheeting, if the sheeting is a material of limited combustibility and both surfaces of the insulating layer have a surface spread of flame of at least Class 0 or 1 (National class) or Class C-s3, d2 or better (European class) (see Appendix A) and make contact with the inner and outer skins of cladding (see Diagram 36) ; or

f. below a floor next to the ground or oversite concrete, if the cavity is less than 1000mm in height or if the cavity is not normally accessible by persons, unless there are openings in the floor such that it is possible for combustibles to accumulate in the cavity (in which case cavity barriers should be provided and access should be provided to the cavity for cleaning).

Note: When a classification includes "s3, d2", this means that there is no limit set for smoke production and/or flaming droplets/particles.

Diagram 34 Cavity wall excluded from provisions for cavity barriers

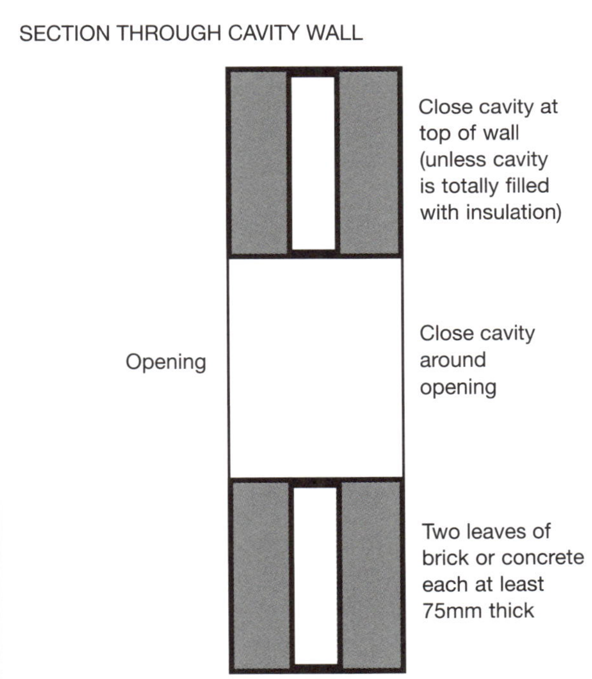

Notes:
1. Domestic meter cupboards may be installed provided that:
   a) there are no more than two cupboards per dwelling.
   b) the openings in the outer wall leaf is not more than 800x500mm for each cupboard.
   c) the inner leaf is not penetrated except by a sleeve not more than 80x80mm, which is fire stopped.
2. Combustible materials may be placed within the cavity.

Diagram 35 Fire-resisting ceiling below concealed space

See para 9.10(c)

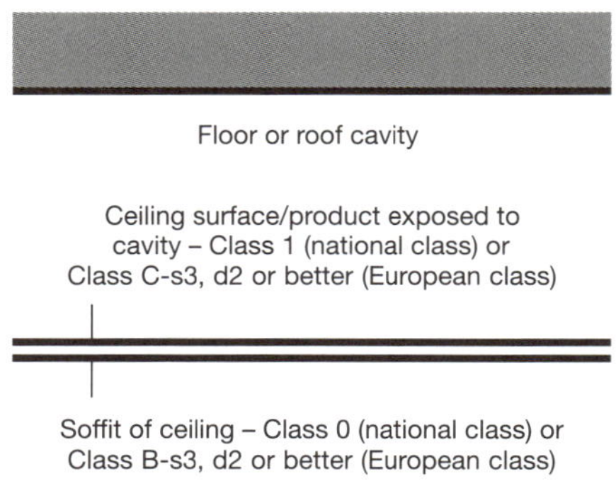

Notes:

1. The ceiling should:
   a. have at least 30 minutes fire resistance;
   b. be imperforate, except for an opening described in paragraph 9.16;
   c. extend throughout the building or compartment; and
   d. not be easily demountable.
2. The National classifications do not automatically equate with the equivalent classifications in the European column, therefore products cannot typically assume a European class unless they have been tested accordingly.
3. When a classification includes "s3, d2", this means that there is no limit set for smoke production and/or flaming droplets/particles.

Diagram 36 Provisions for cavity barriers in double-skinned insulated roof sheeting

See para 9.10(e)

a. ACCEPTABLE WITHOUT CAVITY BARRIERS

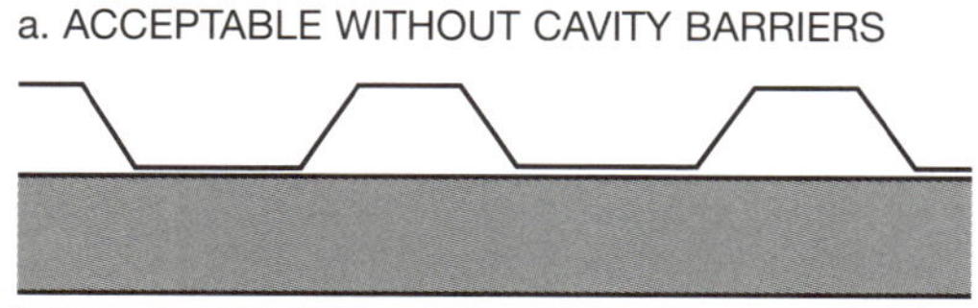

The insulation should make contact with both skins of sheeting. See also Diagram 30a regarding the need for a fire break where such roofs pass over the top of a compartment wall.

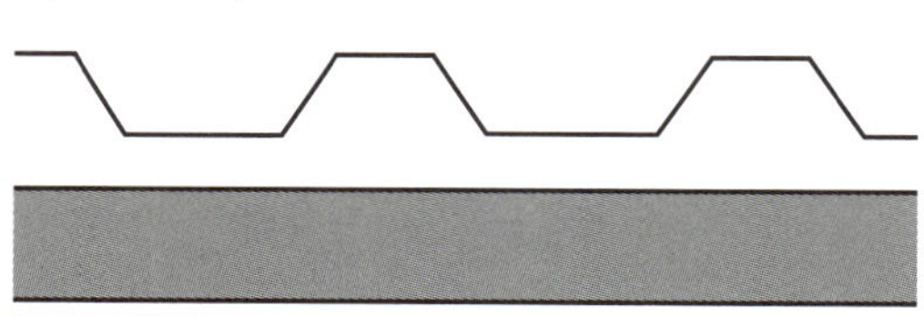

b. CAVITY BARRIERS NECESSARY

9.11 Where any single room with a ceiling cavity or underfloor service void exceeds the dimensions given in Table 13, cavity barriers need only be provided on the line of the enclosing walls/partitions of that room, subject to:

a. the cavity barriers being no more than 40m apart; and

b. the surface of the material/product exposed in the cavity being Class 0 or Class 1 (National class) or Class C-s3, d2 or better (European class).

Note: When a classification includes "s3, d2", this means that there is no limit set for smoke production and/or flaming droplets/particles.

9.12 Where the concealed space is an undivided area which exceeds 40m (this may be in both directions on plan) there is no limit to the size of the cavity if:

a. the room and the cavity together are compartmented from the rest of the building;

b. an automatic fire detection and alarm system meeting the relevant recommendations of BS 5839-1:2002 is fitted in the building. Detectors are only required in the cavity to satisfy BS 5839-1.

c. the cavity is used as a plenum and the recommendations about recirculating air distribution systems in BS 5588-9:1999 are followed;

d. the surface of the material/product used in the construction of the cavity which is exposed in the cavity is Class 0 (National class) or Class B-s3, d2 or better (European class) and the supports and fixings in the cavity are of non-combustible construction;

e. the flame spread rating of any pipe insulation system is Class 1 or Class C-s3, d2 or better (European class) (see Appendix A);

f. any electrical wiring in the void is laid in metal trays, or in metal conduit; and

g. any other materials in the cavity are of limited combustibility or Class A2 or better (European class) (see Appendix A).

Note: When a classification includes "s3, d2", this means that there is no limit set for smoke production and/or flaming droplets/particles.

With respect to cables in communication rooms, typically located below platform floors, because of the substantial number of cables generally involved, with the consequent increase in fire load and where cables are installed in a cavity of 300mm or greater, then the following should be considered:

- **protecting the cavity with an automatic gaseous system (connected to a fire detector and alarm system) or a sprinkler system; or**

- using cables that have been tested and approved to BS EN 50289-4-11:2002 or NFPA 262 or other equivalent specification acceptable to insurers; and
- ensuring that the platform floor has fire resistance of 15 minutes integrity and insulation when exposed to the heating conditions of BS 476-20:1987 or BS EN 1363-1/:1999.

Construction and fixings for cavity barriers

9.13 Every cavity barrier should be constructed to provide at least 30 minutes fire resistance. It may be formed by any construction provided for another purpose if it meets the provisions for cavity barriers (see Appendix A, Table A1, item 15).

Cavity barriers in a stud wall or partition, or provided around openings may be formed of:

a. steel at least 0.5mm thick (this on its own does not provide insulation and should not be used without additional insulation. Such insulation should be non-combustible or of limited combustibility);

b. timber at least 38mm thick;

c. polythene-sleeved mineral wool, or mineral wool slab, in either case under compression when installed in the cavity; or

d. calcium silicate, cement-based or gypsum-based boards at least 12mm thick.

Note: Cavity barriers provided around openings may be formed by the window or door frame if the frame is constructed of steel or timber of the minimum thickness in a) or b) above as appropriate.

9.14 A cavity barrier should, wherever possible, be tightly fitted to a rigid construction and mechanically fixed in position. Where this is not possible (for example, in the case of a junction with slates, tiles, corrugated sheeting or similar materials) the junction should be fire-stopped. Provisions for fire-stopping are set out in Section 10.

9.15 Cavity barriers should also be fixed so that their performance is unlikely to be made ineffective by:

a. movement of the building due to subsidence, shrinkage or temperature change and movement of the external envelope due to wind;

b. collapse in a fire of any services penetrating them;

c. failure in a fire of their fixings (but see note below); and

d. failure in a fire of any material or construction which they abut. (For example, if a suspended ceiling is continued over the top of a fire-resisting wall or partition and direct connection is made between the ceiling and the cavity barrier above the line of the wall or partition, premature failure of the cavity barrier can occur when the ceiling collapses. However, this may not arise if the ceiling is designed to provide fire protection of 30 minutes or more.)

Note: Where cavity barriers are provided in roof spaces, the roof members to which they are fitted are not expected to have any fire resistance – for the purpose of supporting the cavity barrier(s).

9.16 Any openings in a cavity barrier should be limited to those for:

a. doors which have at least 30 minutes fire resistance (see Appendix B, Table B1, item 8) and are fitted in accordance with the provisions of Appendix B;

b. the passage of pipes which meet the provisions in Section 10;

c. the passage of cables or conduits containing one or more cables;

d. openings fitted with a suitably mounted automatic fire damper (see paragraphs 10.11 to 10.15); and

e. ducts which (unless they are fire-resisting) are fitted with a suitably mounted automatic fire damper where they pass through the cavity barrier.

Note: If a cavity barrier is provided above a partition separating bedrooms in accordance with paragraph 9.7 which do not need to be fire resisting partitions then a to e need not apply. However, opinings in the barrier should be kept to a minimum and any penetrations should be sealed to restict the passage of smoke.

Section 10: Protection of openings and fire-stopping

Introduction

10.1 Sections 8 and 9 make provisions for fire-separating elements and set out the circumstances in which there may be openings in them. This section deals with the protection of openings in such elements.

10.2 If a fire-separating element is to be effective, every joint or imperfection of fit, or opening to allow services to pass through the element, should be adequately protected by sealing or fire-stopping so that the fire resistance of the element is not impaired.

10.3 The measures in this Section are intended to delay the passage of fire. They generally have the additional benefit of retarding smoke spread, but the tests specified in Appendix A for integrity does not directly stipulate criteria for the passage of smoke.

10.4 Detailed guidance on door openings and fire doors is given in Appendix B.

Openings for pipes

10.5 Pipes which pass through a fire-separating element (unless the pipe is in a protected shaft), should meet the appropriate provisions in alternatives A, B or C below.

Alternative A: Proprietary seals (any pipe diameter)

10.6 Provide a proprietary sealing system which has been shown by test to maintain the fire resistance of the wall, floor or cavity barrier.

It is important that the test result is applicable to the particular service and material from which the service is made. Potentially different levels of fire resistance performance are possible. Some generic materials often used for this purpose may not be suitable. Ideally, only tested systems should be specified.

Alternative B: Pipes with a restricted diameter

10.7 Where a proprietary sealing system is not used, fire-stopping may be used around the pipe, keeping the opening as small as possible. The nominal internal diameter of the pipe should not be more than the relevant dimension given in Table 14.

Table 14 **Maximum nominal internal diameter of pipes passing through a compartment wall/floor** (see paragraph 10.5 onwards)

Situation		Pipe material and maximum nominal internal diameter (mm)		
		(a) **Non-combustible material** [1]	**(b)** **Lead, aluminium, aluminium alloy, uPVC** [2]**, fibre cement**	**(c)** **Any other material**
1.	Structure (but not a wall separating buildings) enclosing a protected shaft which is not a stairway or a lift shaft	160	110	40
2.	Compartment wall or compartment floor between flats	160	160 (stack pipe) [3] 110 (branch pipe) [3]	40
3.	Any other situation	160	40	40

Notes:

1. Any non-combustible material (such as cast iron, copper or steel) which, if exposed to a temperature of 800°C, will not soften or fracture to the extent that flame or hot gas will pass through the wall of the pipe.
2. uPVC pipes complying with BS 4514:2001 and uPVC pipes complying with BS 5255:1989.
3. These diameters are only in relation to pipes forming part of an above-ground drainage system and enclosed as shown in Diagram 38. In other cases the maximum diameters against situation 3 apply.

Provisions (b) and (c) in this table are only acceptable for property protection if pipework passing through a compartment wall or floor is protected by a suitable pipe closure or seal to close any gaps formed by melting/deforming pipes.

Plastic pipes and small service ducts which pass through a compartment wall, floor or cavity barrier should be fitted with a closing device that is capable of satisfying the integrity criteria (E) for the duration given in Table A1, regardless of the diameter.

The diameters given in Table 14 for pipes of specification (b) used in situation (2) assumes that the pipes are part of an above ground drainage system and are enclosed as shown in Diagram 38 if they are not, the smaller diameter given for situation (3) should be used.

Alternative A above is preferred to Alternative B as some generic materials used for fire stopping may be liable to shrinkage or may not be able to compensate for deflection/movement of the service passing through the wall, floor or cavity barrier. Such factors may lead to lower fire resistance.

Alternative C: Sleeving

10.8 A pipe of lead, aluminium, aluminium alloy, fibre-cement or uPVC, with a maximum nominal internal diameter of 160mm, may be used with a sleeving of non-combustible pipe as shown in Diagram 37. The specification for non-combustible and uPVC pipes is given in the notes to Table 14.

Diagram 37 **Pipes penetrating structure**

See para 10.8, alternative method C

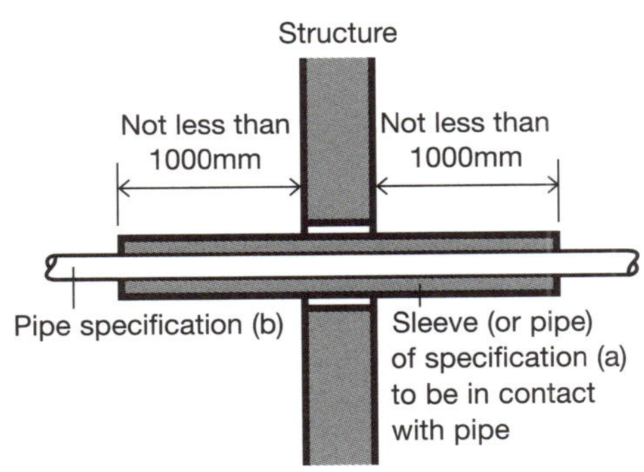

Notes:

1 Make the opening in the structure as small as possible and provide fire-stopping between pipe and structure.

2 See Table 14 for materials specification.

Ventilation ducts, flues etc.

10.9 Where air handling ducts pass through fire separating elements the integrity of those elements should be maintained.

There are three basic methods and these are:

- Method 1 Protection using fire dampers;
- Method 2 Protection using fire-resisting enclosures;
- Method 3 Protection using fire-resisting ductwork.

10.10 Method 1 is not suitable for extract ductwork serving kitchens. This is due to the likely build up of grease within the duct which can adversely affect the effectiveness of any dampers.

Build-up of grease in kitchen extract ducts can be reduced by specifying effective grease filters, for example, those tested and approved to LPS 1263. All filters must be kept clean to further reduce the risk of transfer of grease into the duct. In addition, adequate access doors must be provided to enable cleaning of the inside surfaces of the duct.

Further information on fire-resisting ductwork is given in the ASFP Blue Book: *Fire resistng ductwork* (ISBN: 1 87040 926 4) published by the Association for Specialist Fire Protection and freely available from the ASFP website at www.asfp.org.uk.

Fire dampers

10.11 Fire dampers should be situated within the thickness of the fire-separating elements and be securely fixed. It is also necessary to ensure that, in a fire, expansion of the ductwork would not push the fire damper through the structure.

10.12 Adequate means of access should be provided to allow inspection, testing and maintenance of both the fire damper and its actuating mechanism.

10.13 Where the use of the building involves a sleeping risk, such as an hotel or residential care home, fire dampers should be actuated by smoke detector-controlled automatic release mechanisms, in addition to being actuated by thermally actuated devices.

However, in a situation where all occupants of the building can be expected to make an unaided escape and an L1 fire alarm system is installed in accordance with BS 5839-1:2002, the following exceptions may be made:

a. If, on the detection of smoke, the alarm system signals the immediate evacuation of all the occupants of the building, then fire/smoke dampers are not needed; and

b. If the building is divided into fire compartments and the alarm system is arranged to signal the immediate evacuation of the occupants of the fire compartment in which the fire has been detected, then smoke detector operated fire/smoke dampers need only be provided where ductwork enters or leaves the fire compartment.

Note: Fire dampers actuated only by fusible links are not suitable for protecting escape routes (or areas of special risk). However an ES classified fire and smoke damper which is activated by a suitable fire detection system may be used. See paragraph 10.15.

10.14 Further guidance on the design and installation of mechanical ventilation and air-conditioning plant is given in BS 5720:1979 on ventilation and air-conditioning ductwork in BS 5588-9:1999.

Further information on fire and smoke-resisting dampers is given in the ASFP Grey Book: *Fire and smoke resisting dampers'* (ISBN: 1 87040 924 8) published by the Association for Specialist Fire Protection and freely available from the ASFP website at www.asfp.org.uk.

10.15 Fire dampers should be tested to BS EN 1366-2:1999 and be classified to BS EN 13501-3:2005. They should have an E classification equal to, or greater than, 60 minutes. Fire and smoke dampers should also be tested to BS EN 1366-2:1999 and be classified to BS EN 13501-3. They should have an ES classification equal to, or greater than, 60 minutes.

Note 1: Fire dampers tested using ad-hoc procedures based on BS 476 may only be appropriate for fan-off situations. In all cases, fire dampers should be installed as tested.

Note 2: Paragraphs 5.46 and 8.40 also deal with ventilation and air-conditioning ducts.

Flues, etc.

10.16 If a flue, or duct containing flues or appliance ventilation duct(s), passes through a compartment wall or compartment floor, or is built into a compartment wall, each wall of the flue or duct should have a fire resistance of at least half that of the wall or floor in order to prevent the by-passing of the compartmentation (see Diagram 39).

Fire-stopping

10.17 In addition to any other provisions in this document for fire-stopping:

a. joints between fire-separating elements should be fire-stopped; and

b. all openings for pipes, ducts, conduits or cables to pass through any part of a fire-separating element should be:

    i. kept as few in number as possible; and

    ii. kept as small as practicable; and

    iii. fire-stopped (which in the case of a pipe or duct, should allow thermal movement).

10.18 To prevent displacement, materials used for fire-stopping should be reinforced with (or supported by) materials of limited combustibility in the following circumstances:

a. in all cases where the unsupported span is greater than 100mm; and

b. in any other case where non-rigid materials are used (unless they have been shown to be satisfactory by test).

10.19 Proprietary fire-stopping and sealing systems (including those designed for service penetrations) which have been shown by test to maintain the fire resistance of the wall or other element, are available and may be used.

Other fire-stopping materials include:

- cement mortar;
- gypsum-based plaster;
- cement-based or gypsum-based vermiculite/perlite mixes;
- glass fibre, crushed rock, blast furnace slag or ceramic-based products (with or without resin binders); and
- intumescent mastics.

These may be used in situations appropriate to the particular material. Not all of them will be suitable in every situation.

The following is included for guidance for property protection purposes.

When a fire separating element, including a large cavity barrier, is penetrated by a service, the fire resistance of the element shall be reinstated to the levels recommended in Table A1 in Appendix A as appropriate, by one of the following methods:

a. Small single pipes and cables (not exceeding 12mm diameter):

i. where the hole through which the service passes is no greater than 2 x the diameter of the service, the resultant gap shall be fire-stopped to the full depth of the barrier.

ii. where the hole is greater than 2 x the diameter of the service, the resultant gap shall be sealed with a proprietary sealant capable of satisfying the integrity and insulation criteria for the duration given in Table A1.

b. Large metal pipes, large cables or bunches of unsupported cables passing through a small penetration.

i. the gap between the service and the seal shall be sealed with a proprietary sealant capable of satisfying the integrity (E) and insulation (I) criteria for the duration recommended in Table A1.

In addition, the surface of the service shall not exhibit a temperature rise in excess of 250°C for a distance of not less than 500mm on the non-fire exposed side of the barrier.

c. Large metal pipes, groups of pipes, large cables, large bunches of cables, with or without support from ladders or trays passing through large penetrations (see definition).

i. the aperture in the separating construction shall be filled by a bulkhead barrier of adequate robustness which is able to satisfy integrity (E) and insulation (I) criteria for the durations given in Table A1, when tested in accordance with the referenced test. Gaps between the service and the bulkhead barriers shall be sealed by a proprietary sealant capable of satisfying the integrity and insulation criteria for the same duration. The surface of any service or supporting frame shall not exceed a temperature rise in excess of 250°C for a distance of not less than 500mm on the non-fire exposed side of the barrier.

Further information can be found in the IFSA code.

Guidance on the process of design, installation and maintenance of passive fire protection is available in *Ensuring best practice for passive fire protection in buildings* (ISBN: 1 87040 919 1) produced by the Association for Specialist Fire Protection (ASFP).

Further information on the generic types of systems available, information about their suitability for different applications and guidance on test methods is given in the ASFP Red Book: *Fire Stopping and Penetration Seals for the Construction Industry – the 'Red Book'* (ISBN: 1 87040 923 X) published by the Association for Specialist Fire Protection and freely available from the ASFP website at www.asfp.org.uk.

Diagram 38 **Enclosure for drainage or water supply pipes**

See para 5.24

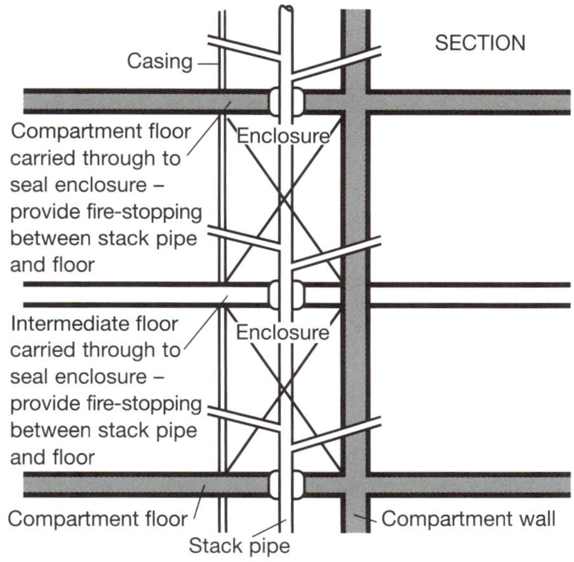

Notes:

1. The enclosure should:
   a. be bounded by a compartment wall or floor, an outside wall, an intermediate floor, or a casing (see specification at 2 below);
   b. have internal surfaces (except framing members) of Class 0 (National class) or Class B-s3, d2 or better (European class)
   Note: When a classification includes 's3, d2', this means that there is no limit set for smoke production and/or flaming droplets/particles);
   c. not have an access panel which opens into a circulation space or bedroom;
   d. be used only for drainage, or water supply, or vent pipes for a drainage system.
2. The casing should:
   a. be imperforate except for an opening for a pipe or an access panel;
   b. not be of sheet metal;
   c. have (including any access panel) not less than 30 minutes fire resistance.
3. The opening for a pipe, either in the structure or the casing, should be as small as possible and fire-stopped around the pipe.

Diagram 39 **Flues penetrating compartment walls or floors**
(note that there is guidance in Approved Document J concerning hearths adjacent to compartment walls)

See para 10.16

a. FLUE PASSING THROUGH COMPARTMENT WALL OR FLOOR

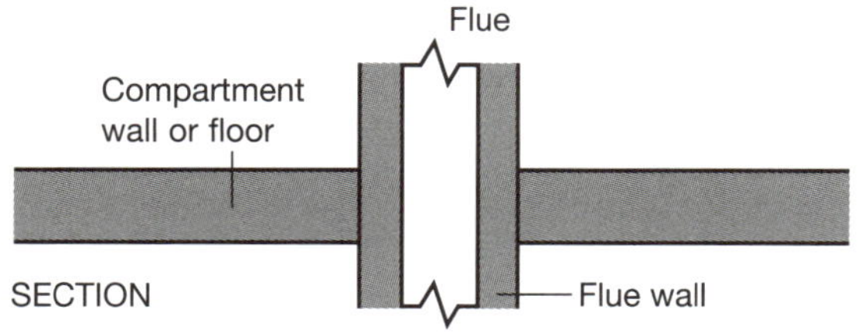

Flue walls should have a fire resistance of at least one half of that required for the compartment wall or floor, and be of non-combustible construction.

b. FLUE BUILT INTO COMPARTMENT WALL

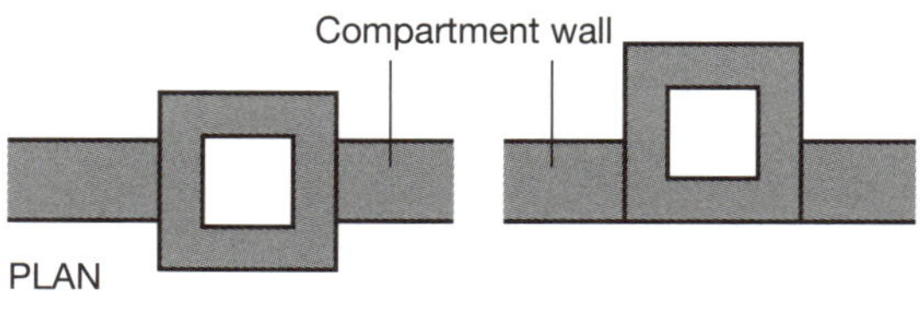

In each case flue walls should have a fire resistance at least one half of that required for the compartment wall and be of non-combustible construction.

Section 11: Special provisions for car parks and shopping complexes

Introduction

11.1 This section describes additional considerations which apply to the design and construction of car parks and shopping complexes.

Car parks

General principles

11.2 Buildings or parts of buildings used as parking for cars and other light vehicles are unlike other buildings in certain respects which merit some departures from the usual measures to restrict fire spread within buildings. Those are:

a. The fire load is well defined; and

b. Where the car park is well ventilated, there is a low probability of fire spread from one storey to another. Ventilation is the important factor and, as heat and smoke cannot be dissipated so readily from a car park that is not open-sided, fewer concessions are made. The guidance in paragraphs 11.3 to 11.6 is concerned with three ventilation methods: open-sided (high level of natural ventilation), natural ventilation and mechanical ventilation.

Open-sided car parks

11.3 If the building, or separated part containing the car park, complies with the following provisions it may be regarded as an open-sided car park for the purposes of fire resistance assessment in Appendix A, Table A2. The provisions are that:

a. there should not be any basement storeys;

b. each storey should be naturally ventilated by permanent openings at each car parking level, having an aggregate vent area not less than 1/20th of the floor area at that level, of which at least half (1/40th) should be equally provided between two opposing walls;

c. where one element of structure supports or carries or gives stability to another, the fire resistance of the supporting element should be no less than the minimum period of fire resistance for the other element (whether that other element is loadbearing or not).

d. if the building is also used for any other purpose, the part forming the car park is a separated part and the fire resistance of any element of structure that supports or carries or gives stability to another element in the other part of the building should be no less than the minimum period of fire resistance for the elements it supports; and

e. all materials used in the construction of the building, compartment or separated part should be non-combustible, except for:

    i. any surface finish applied to a floor or roof of the car park, or within any adjoining building, compartment or separated part to the structure enclosing the car park, if the finish meets all relevant aspects of the guidance on requirements B2 and B4;

    ii. any fire door;

    iii. any attendant's kiosk not exceeding 15m^2 in area; and

    iv. any shop mobility facility.

Car parks which are not open-sided

11.4 Where car parks do not have the standard of ventilation set out in paragraph 11.3(b), they are not regarded as open-sided and a different standard of fire resistance is necessary (the relevant provisions are given in Appendix A, Table A2).

Such car parks still require some ventilation, which may be by natural or mechanical means, as described in paragraphs 11.5 or 11.6 below.

Natural ventilation

11.5 Where car parks that are not open-sided are provided with some, more limited, natural ventilation, each storey should be ventilated by permanent openings (which can be at ceiling level) at each car parking level. These should have an aggregate free vent area not less than 1/40th of the floor area at that level, of which at least half should be split equally and provided between two opposing walls (1/160th on each side). (See Approved Document F *Ventilation* for additional guidance on normal ventilation of car parks.)

Mechanical ventilation

11.6 In most basement car parks and in enclosed car parks, it may not be possible to obtain the minimum standard of natural ventilation openings set out in paragraph 11.5. In such cases a system of mechanical ventilation should be provided as follows:

a. the system should be independent of any other ventilating system (other than any system providing normal ventilation to the car park) and be designed to operate at 10 air changes per hour in a fire condition. (See Approved Document F *Ventilation* for guidance on normal ventilation of car parks);

b. the system should be designed to run in two parts, each part capable of extracting 50% of the rates set out in (a) above and designed so that each part may operate singly or simultaneously;

c. each part of the system should have an independent power supply which would operate in the event of failure of the main supply;

d. extract points should be arranged so that 50% of the outlets are at high level and 50% at low level; and

e. the fans should be rated to run at 300°C for a minimum of 60 minutes and the ductwork and fixings should be constructed of materials having a melting point not less than 800°C.

For further information on equipment for removing hot smoke, refer to BS EN 12101-3:2002.

An alternative method of providing smoke ventilation from enclosed car parks is given BS 7346-7:2006.

Shopping complexes

11.7 Whilst the provisions in this document about shops should generally be capable of application in cases where a shop is contained in a single separate building, the provisions may not be appropriate where a shop forms part of a complex. These may include covered malls providing access to a number of shops and common servicing areas. In particular, the provisions about maximum compartment size may be difficult to meet, bearing in mind that it would generally not be practical to compartment a shop from a mall serving it. To a lesser extent, the provisions about fire resistance, walls separating shop units, surfaces and boundary distances may pose problems.

11.8 To ensure a satisfactory standard of fire safety in shopping complexes, alternative measures and additional compensatory features to those set out in this document are appropriate. Such features are set out in Sections 5 and 6 of BS 5588-10:1991 and the relevant recommendations of those sections should be followed.

The Requirement

This Approved Document deals with the following Requirement from Part B of Schedule 1 to the Building Regulations 2000 (as amended).

Requirement	*Limits on application*
External fire spread **B4.** (1) The external walls of the building shall adequately resist the spread of fire over the walls and from one building to another, having regard to the height, use and position of the building. (2) The roof of the building shall adequately resist the spread of fire over the roof and from one building to another, having regard to the use and position of the building.	

Insurers' Essential Principles

The relevant essential principles for property protection are:

- There shall be adequate provision to prevent an arson attack;
- The building shall be so constructed that fire cannot spread into the premises from an adjoining building or other external fire source.

Guidance

Performance

In the Secretary of State's view the Requirements of B4 will be met:

a. if the external walls are constructed so that the risk of ignition from an external source and the spread of fire over their surfaces, is restricted, by making provision for them to have low rates of heat release;

b. if the amount of unprotected area in the side of the building is restricted so as to limit the amount of thermal radiation that can pass through the wall, taking the distance between the wall and the boundary into account; and

c. if the roof is constructed so that the risk of spread of flame and/or fire penetration from an external fire source is restricted.

In each case so as to limit the risk of a fire spreading from the building to a building beyond the boundary, or vice versa.

The extent to which this is necessary is dependent on the use of the building, its distance from the boundary and, in some cases, its height.

Introduction

External walls

B4.i The construction of external walls and the separation between buildings to prevent external fire spread are closely related.

The chances of fire spreading across an open space between buildings and the consequences if it does, depend on:

a. the size and intensity of the fire in the building concerned;

b. the distance between the buildings;

c. the fire protection given by their facing sides; and

d. the risk presented to people in the other building(s).

B4.ii Provisions are made in Section 12 for the fire resistance of external walls and to limit the susceptibility of the external surface of walls to ignition and to fire spread.

B4.iii Provisions are made in Section 13 to limit the extent of openings and other unprotected areas in external walls in order to reduce the risk of fire spread by radiation.

Roofs

B4.iv Provisions are made in Section 14 for reducing the risk of fire spread between roofs and over the surfaces of roofs.

Section 12: Construction of external walls

Introduction

12.1 Provisions are made in this Section for the external walls of the building to have sufficient fire resistance to prevent fire spread across the relevant boundary. The provisions are closely linked with those for space separation in Section 13 which sets out limits on the amount of unprotected area of wall. As the limits depend on the distance of the wall from the relevant boundary, it is possible for some or all of the walls to have no fire resistance, except for any parts which are loadbearing (see paragraph B3.ii).

External walls are elements of structure and the relevant period of fire resistance (specified in Appendix A) depends on the use, height and size of the building concerned. If the wall is 1000mm or more from the relevant boundary, a reduced standard of fire resistance is accepted in most cases and the wall only needs fire resistance from the inside.

12.2 Provisions are also made to restrict the combustibility of external walls of buildings that are less than 1000mm from the relevant boundary and, irrespective of boundary distance, the external walls of high buildings and those of the Assembly and Recreation Purpose Groups. This is in order to reduce the surface's susceptibility to ignition from an external source and to reduce the danger from fire spread up the external face of the building.

In the guidance to Requirement B3, provisions are made in Section 7 for internal and external loadbearing walls to maintain their loadbearing function in the event of fire.

Approved documents supporting national Building Regulations make limited provisions relating to the resistance to fire from penetration from an external fire. Where an external wall is more than 1m from a relevant boundary, there are usually no provisions relating to the resistance to fire penetration from an external fire.

Consequently, under most circumstances external walls only have to have a degree of resistance to fire penetration from an external fire. Doors and windows in an external wall are not counted as part of the wall itself. Rather, doors and windows are defined as 'unprotected area', and do not need to provide any resistance to fire penetration from an internal or external fire. The greater the distance an external wall is from a 'relevant boundary', the more of the 'wall' is allowed to be replaced with 'unprotected area'.

Consequently, from a property protection perspective, in order to meet the essential principle that 'There shall be adequate provision to prevent an arson attack', it is recommended that the fire resistance of an external wall will be considered satisfactory if:

- *there are no 'unprotected openings' within the protected zone; and*
- *the wall is fire-resisting from the outside to the standard described in Table A9.*

Fire resistance standard

12.3 The external walls of the building should have the appropriate fire resistance given in Appendix A, Table A1, unless they form an unprotected area under the provisions of Section 13.

Portal frames

12.4 Portal frames are often used in single storey industrial and commercial buildings where there may be no need for fire resistance of the structure (Requirement B3). However, where a portal framed building is near a relevant boundary, the external wall near the boundary may need fire resistance to restrict the spread of fire between buildings.

It is generally accepted that a portal frame acts as a single structural element because of the moment-resisting connections used, especially at the column/rafter joints. Thus, in cases where the external wall of the building cannot be wholly unprotected, the rafter members of the frame, as well as the column members, may need to be fire protected.

Following an investigation of the behaviour of steel portal frames in fire, it is considered technically and economically feasible to design the foundation and its connection to the portal frame so that it would transmit the overturning moment caused by the collapse, in a fire, of unprotected rafters, purlins and some roof cladding, while allowing the external wall to continue to perform its structural function. The design method for this is set out in the SCI publication *P313 Single storey steel framed buildings in fire boundary conditions,* 2002 (ISBN: 1 85942 135 0).

Note 1: The recommendations in the SCI publication for designing the foundation to resist overturning need not be followed if the building is fitted with a sprinkler system in accordance with paragraph 0.16.

Note 2: Normally, portal frames of reinforced concrete can support external walls requiring a similar degree of fire resistance without specific provision at the base to resist overturning.

Note 3: Existing buildings may have been designed to the following guidance which is also acceptable:

a. the column members are fixed rigidly to a base of sufficient size and depth to resist overturning;

b. there is brick, block or concrete protection to the columns up to a protected ring beam providing lateral support; and

c. there is some form of roof venting to give early heat release. (The roof venting could be, for example, PVC rooflights covering some 10 per cent of the floor area and evenly spaced over the floor area.)

External wall construction

12.5 The external envelope of a building should not provide a medium for fire spread if it is likely to be a risk to health or safety. The use of combustible materials in the cladding system and extensive cavities may present such a risk in tall buildings.

Externall walls should either meet the guidance given in paragraphs 12.6 to 12.9 or meet the performance criteria given in the BRE Report *Fire performance of external thermal insulation for walls of multi storey buildings* (BR 135) for cladding systems using full scale test data from BS 8414-1:2002 or BS 8414-2:2005.

The total amount of combustible material may also be limited in practice by the provisions for space separation in Section 13 (see paragraph 13.7 onwards).

External surfaces

12.6 The external surfaces of walls should meet the provisions in Diagram 40. Where a mixed use building includes Assembly and Recreation Purpose Group(s) accommodation, the external surfaces of walls should meet the provisions in Diagram 40c.

Insulation Materials/Products

12.7 In a building with a storey 18m or more above ground level any insulation product, filler material (not including gaskets, sealants and similar) etc. used in the external wall construction should be of limited combustibility (see Appendix A). This restriction does not apply to masonry cavity wall construction which complies with Diagram 34 in Section 9.

Cavity barriers

12.8 Cavity barriers should be provided in accordance with Section 9.

12.9 In the case of a an external wall construction, of a building which, by virtue of paragraph 9.10d (external cladding system with a masonry or concrete inner leaf), is not subject to the provisions of Table 13 *Maximum dimensions of cavities in non-domestic buildings*, the surfaces which face into cavities should also meet the provisions of Diagram 40.

Diagram 40 **Provisions for external surfaces or walls**

See paras 12.5 and 12.6

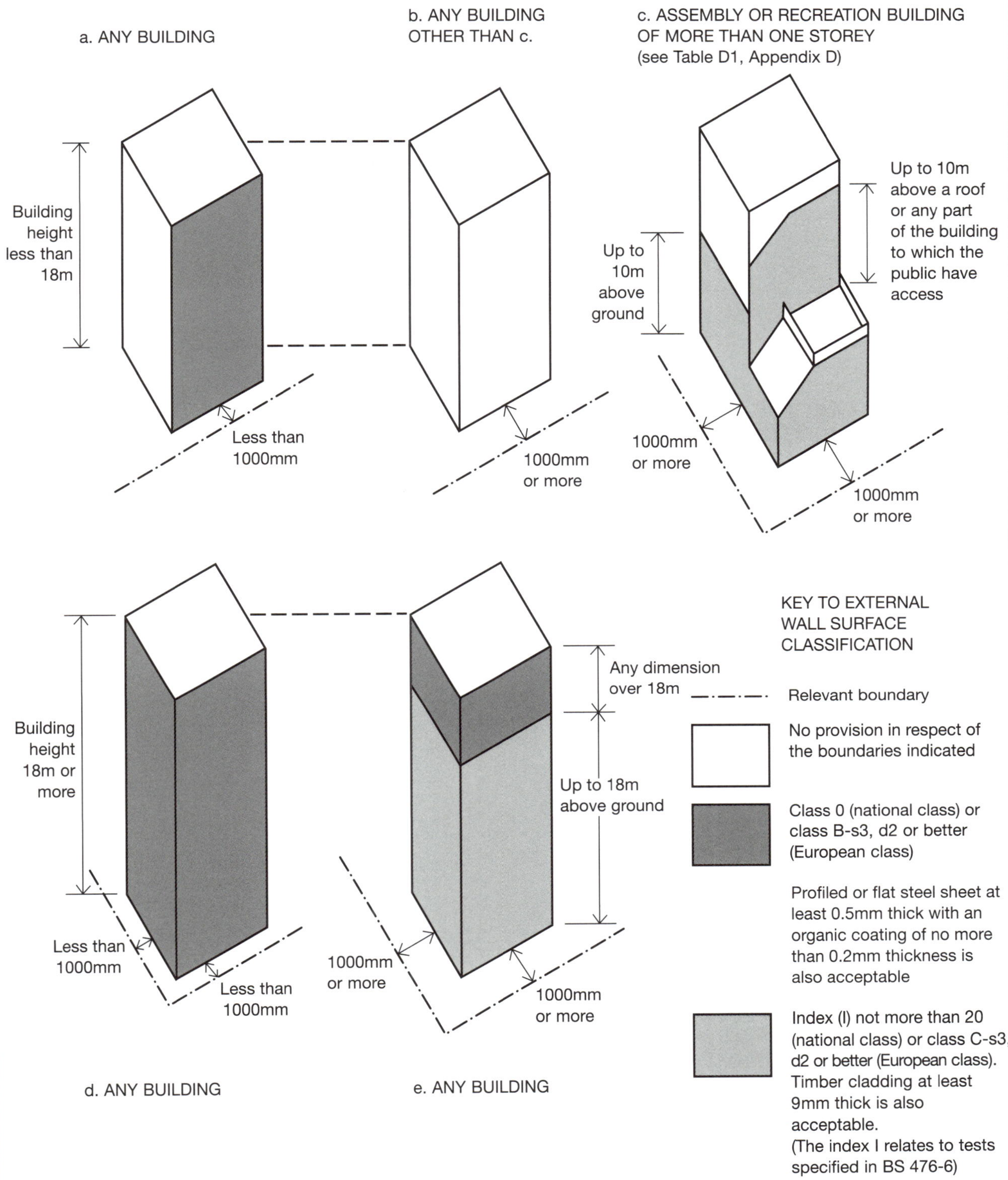

Notes:

1 The national classifications do not automatically equate with the equivalent European classifications, therefore, products cannot typically assume a European class unless they have been tested accordingly.

2 When a classification includes "s3, d2", this means that there is no limit set for smoke production and/or flaming droplets/particles.

Section 13: Space separation

Introduction

13.1 The provisions in this Section are based on a number of assumptions and, whilst some of these may differ from the circumstances of a particular case, together they enable a reasonable standard of space separation to be specified. The provisions limit the extent of unprotected areas in the sides of a building (such as openings and areas with a combustible surface) which will not give adequate protection against the external spread of fire from one building to another.

A roof is not subject to the provisions in this Section unless it is pitched at an angle greater than 70° to the horizontal (see definition for 'external wall' in Appendix E). Similarly, vertical parts of a pitched roof such as dormer windows (which taken in isolation might be regarded as a wall), would not need to meet the following provisions unless the slope of the roof exceeds 70°. It is a matter of judgement whether a continuous run of dormer windows occupying most of a steeply pitched roof should be treated as a wall rather than a roof.

13.2 The assumptions are:

a. that the size of a fire will depend on the compartmentation of the building, so that a fire may involve a complete compartment, but will not spread to other compartments;

b. that the intensity of the fire is related to the use of the building (i.e. purpose group), but that it can be moderated by a sprinkler system;

c. that Residential and Assembly and Recreation Purpose Groups represent a greater life risk than other uses;

d. that there is a building on the far side of the boundary that has a similar elevation to the one in question and that it is at the same distance from the common boundary; and

e. that the amount of radiation that passes through any part of the external wall that has fire resistance may be discounted.

13.3 Where a reduced separation distance is desired (or an increased amount of unprotected area) it may be advantageous to construct compartments of a smaller size.

Diagram 41 **Relevant boundary**

See paras 13.4 and 13.5

This diagram sets out the rules that apply in respect of a boundary for it to be considered as a relevant boundary.

For a boundary to be relevant it should:

a coincide with; or
b be parallel to; or
c be at an angle of not more than 80° to the side of the building

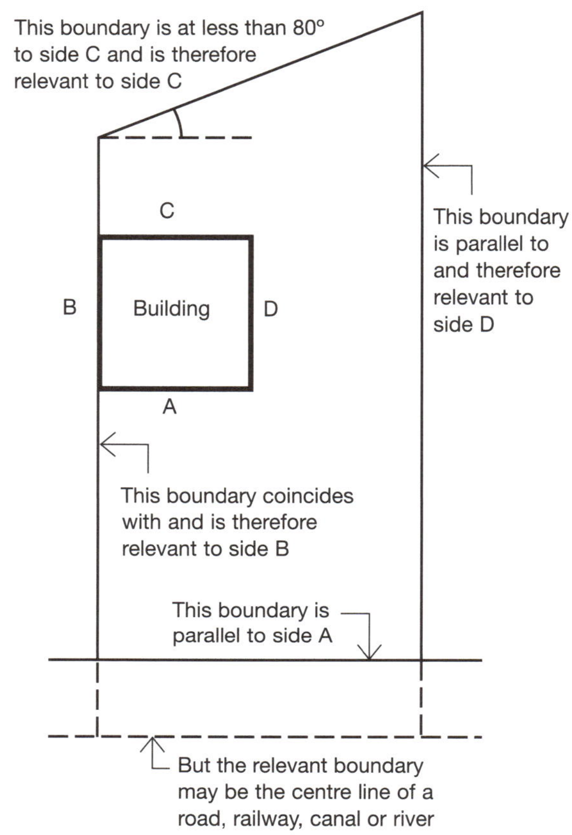

Boundaries

13.4 The use of the distance to a boundary, rather than to another building, in measuring the separation distance, makes it possible to calculate the allowable proportion of unprotected areas, regardless of whether there is a building on an adjoining site and regardless of the site of that building or the extent of any unprotected areas that it might have.

A wall is treated as facing a boundary if it makes an angle with it of 80° or less (see Diagram 41).

Usually only the distance to the actual boundary of the site needs to be considered. But in some circumstances, when the site boundary adjoins a space where further development is unlikely, such as a road, then part of the adjoining space may be included as falling within the relevant boundary for the purposes of this Section. The meaning of the term boundary is explained in Diagram 41.

Relevant boundaries

13.5 The boundary which a wall faces, whether it is the actual boundary of the site or a notional boundary, is called the relevant boundary (see Diagrams 41 and 42).

Notional boundaries

13.6 Generally separation distance between buildings on the same site is discounted. In some circumstances the distances to other buildings on the same site need to be considered. This is done by assuming that there is a boundary between those buildings. This assumed boundary is called a notional boundary.

A notional boundary is assumed to exist where:

a. either or both of the buildings concerned are in the Residential or Assembly and Recreation Purpose Groups; or

b. more than one building is constructed on the same site but is to be operated/managed by different organisations.

The appropriate rules are given in Diagram 42. Considerations for property protection issues are shown in Diagram 42a.

Diagram 42 **Notional boundary**

See para 13.6

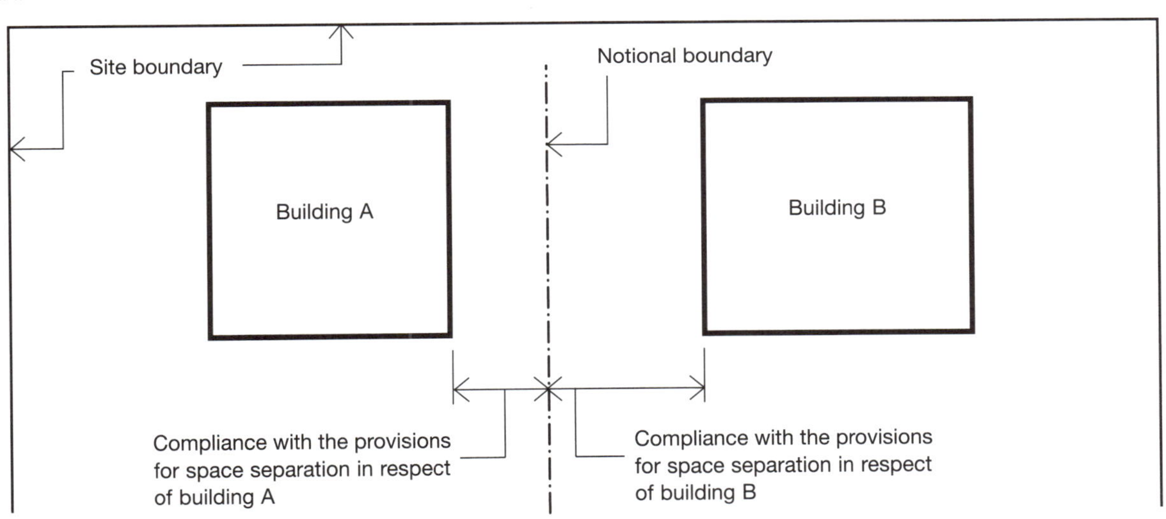

The notional boundary should be set in the area between the two buildings using the following rules:

1. The notional boundary is assumed to exist in the space between the buildings and is positioned so that one of the buildings would comply with the provisions for space separation having regard to the amount of its unprotected area. In practice, if one of the buildings is existing, the position of the boundary will be set by the space separation factors for that building.
2. The siting of the new building, or the second building if both are new, can then be checked to see that it also complies, using the notional boundary as the relevant boundary for the second building.

Diagram 42a Combustible goods stored between building and notional boundary

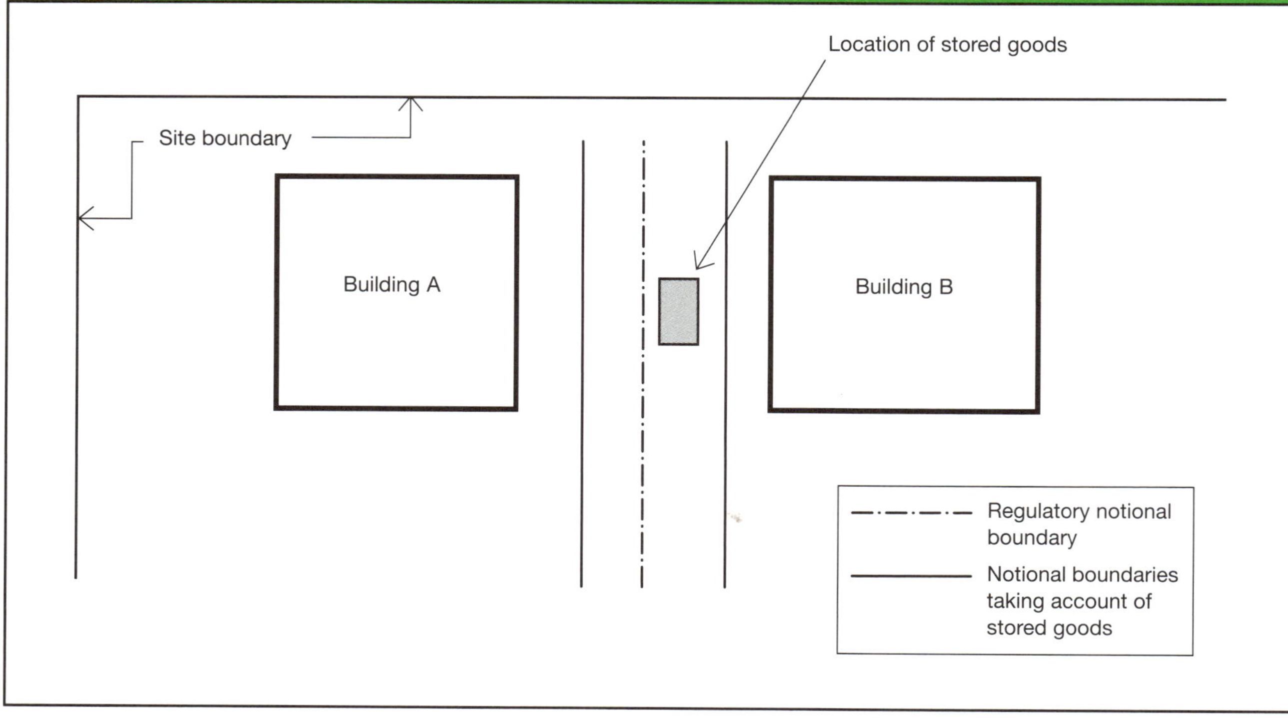

This diagram illustrates the concern that insurers have when two buildings have combustible materials stored between the building and the notional boundary specified by regulators.

The basic concept is that the location of combustible goods relocates the notional boundary closer to the buildings and consequently higher performance requirements for the external walls need to be specified. The effect is similar to moving two buildings closer to each other, although the area of exposure may not be the same.

Whilst the use of a notional boundary ensures as far as possible that the possibility of fire spread from one building to another is minimised, it fails to take into account the effect of any combustible materials such as pallets stored between the buildings and consequently higher levels of fire resistance may be appropriate (see Appendix A, Table A9 for guidance).

Unprotected areas and fire resistance

13.7 Any part of an external wall which has less fire resistance than the appropriate amount given in Appendix A, Table A2, is considered to be an unprotected area.

External walls of protected shafts forming stairways

13.8 Any part of an external wall of a stairway in a protected shaft is excluded from the assessment of unprotected area.

Note: There are provisions in the guidance to B1 (Diagram 24) and B5 (paragraph 17.11) which refers to Section 2 of BS 5588-5:2004 about the relationship of external walls for protected stairways to the unprotected areas of other parts of the building.

Status of combustible surface materials as unprotected area

13.9 If an external wall has the appropriate fire resistance, but has combustible material more than 1mm thick as its external surface, then that wall is counted as an unprotected area amounting to half the actual area of the combustible material, see Diagram 43. (For the purposes of this provision, a material with a Class 0 rating (National class) or Class B-s3, d2 rating (European class) (see Appendix A, paragraphs 7 and 13) need not be counted as unprotected area).

Note: When a classification includes "s3, d2", this means that there is no limit set for smoke production and/or flaming droplets/particles.

Diagram 43 **Status of combustible surface material as unprotected area**

See para 13.9

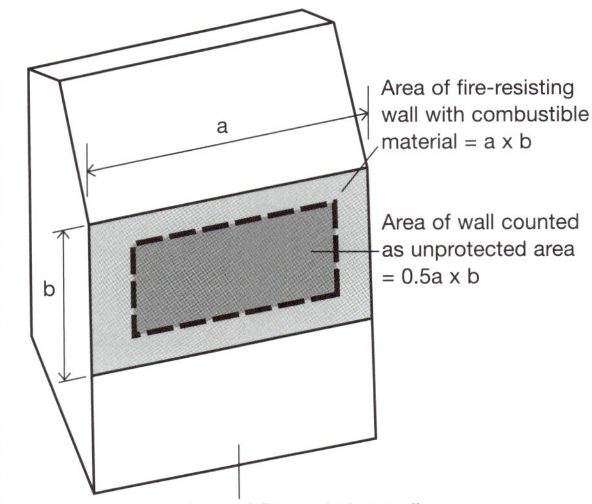

Diagram 43a **Property protection provisions for external surfaces of walls**

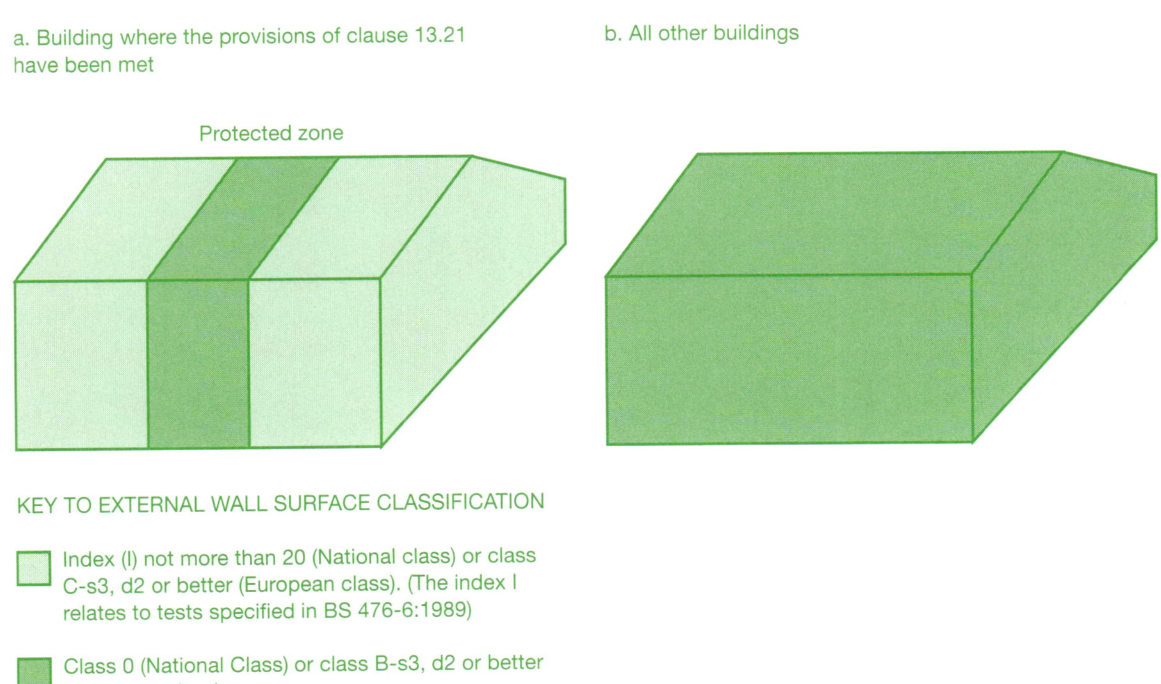

Small unprotected areas

13.10 Small unprotected areas in an otherwise protected area of wall are considered to pose a negligible risk of fire spread and may be disregarded. Diagram 44 shows the constraints that apply to the placing of such areas in relation to each other and to lines of compartmentation inside the building. These constraints vary according to the size of each unprotected area.

Canopies

13.11 Some canopy structures would be exempt from the application of the Building Regulations by falling within Class VI or Class VII of Schedule 2 to the Regulations (Exempt buildings and works). Many others may not meet the exemption criteria and in such cases the provisions in this Section about limits of unprotected areas could be onerous.

In the case of a canopy attached to the side of a building, provided that the edges of the canopy are at least 2m from the relevant boundary, separation distance may be determined from the wall rather than the edge of the canopy (see Diagram 45).

In the case of a free-standing canopy structure above a limited risk or controlled hazard (for example over petrol pumps), in view of the high degree of ventilation and heat dissipation achieved by the open sided construction and provided the canopy is 1000mm or more from the relevant boundary, the provisions for space separation could reasonably be disregarded.

Large uncompartmented buildings

13.12 Parts of the external wall of an uncompartmented building which are more than 30m above mean ground level, may be disregarded in the assessment of unprotected area.

Diagram 44 **Unprotected areas which may be disregarded in assessing the separation distance from the boundary**

See para 13.10

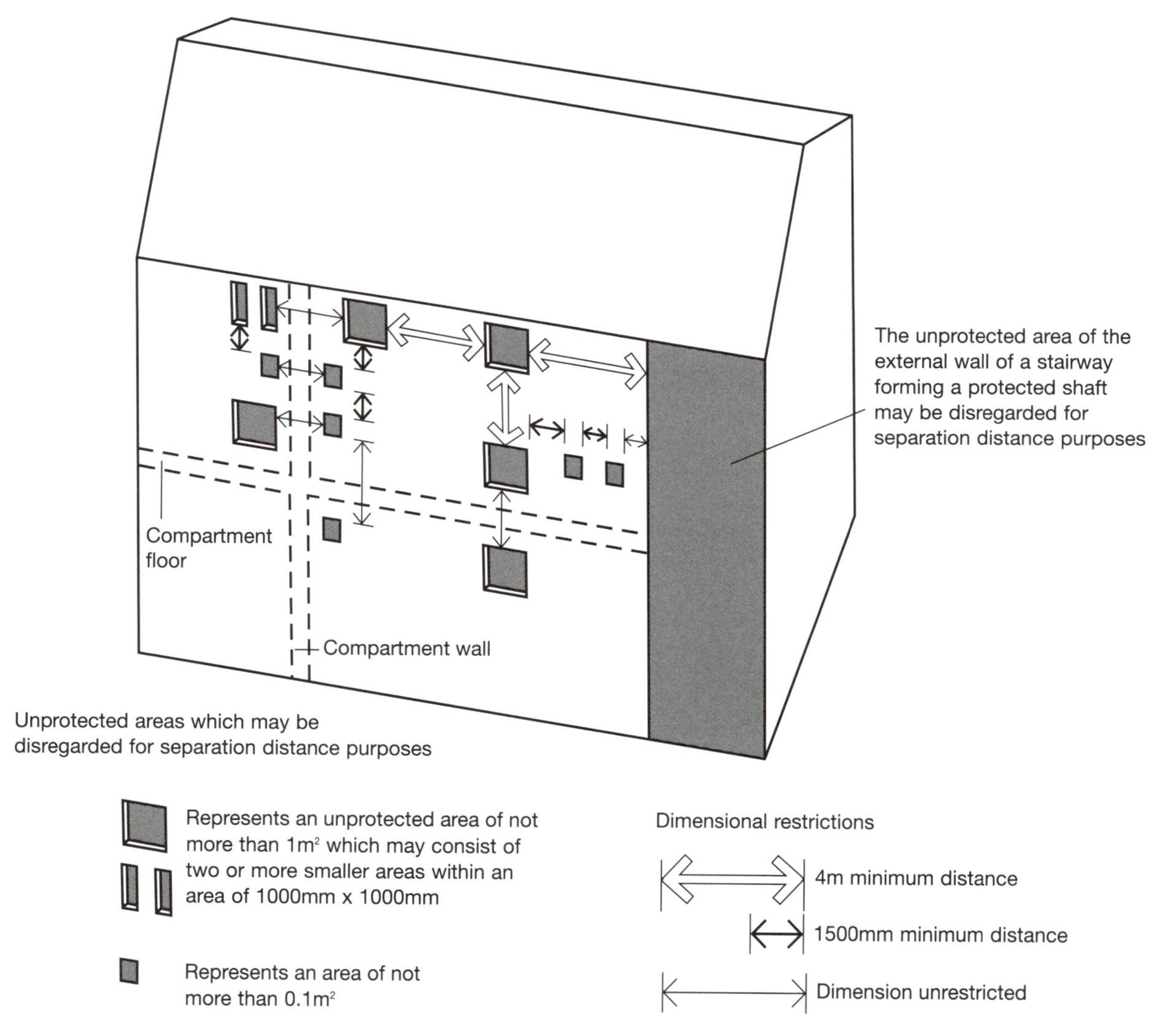

Diagram 45 **The effect of a canopy on separation distance**

See para 13.11

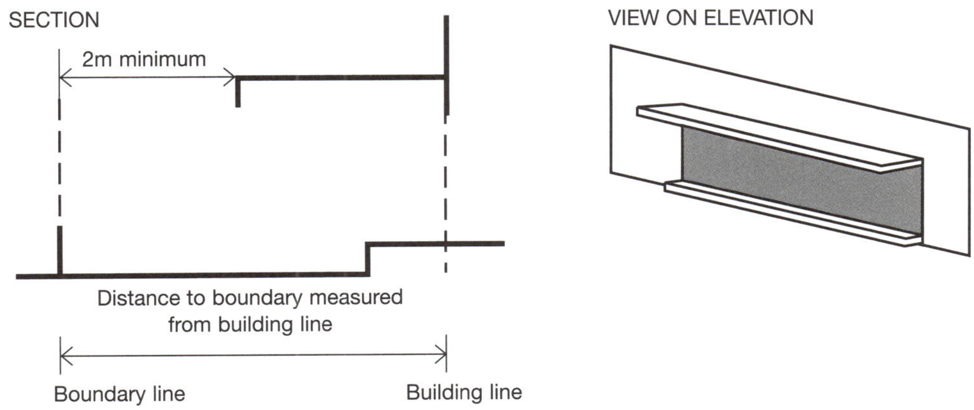

Projections from the building line such as a canopy or a loading platform can be ignored when assessing separation distance. This would not apply to an enclosed loading bay, for example if the illustration had shown side walls beneath the canopy.

External walls within 1000mm of the relevant boundary

13.13 A wall situated within 1000mm from any point on the relevant boundary and including a wall coincident with the boundary, will meet the provisions for space separation if:

a. the only unprotected areas are those shown in Diagram 44 or referred to in paragraph 13.12; and

b. the rest of the wall is fire-resisting from both sides.

External walls 1000mm or more from the relevant boundary

13.14 A wall situated at least 1000mm from any point on the relevant boundary will meet the provisions for space separation if:

a. the extent of unprotected area does not exceed that given by one of the methods referred to in paragraph 13.15; and

b. the rest of the wall (if any) is fire-resisting from the inside of the building.

Methods for calculating acceptable unprotected area

13.15 Two simple methods are given in this Approved Document for calculating the acceptable amount of unprotected area in an external wall that is at least 1000mm from any point on the relevant boundary. (For walls within 1000mm of the boundary see 13.13 above.)

Method 1 may be used for small residential buildings which do not belong to Purpose Group 2a (Institutional type premises) and is set out in paragraph 13.19.

Method 2 may be used for most buildings or compartments for which Method 1 is not appropriate and is set out in paragraph 13.20.

There are other more precise methods, described in a BRE report *External fire spread: Building separation and boundary distances* (BR 187, BRE 1991), which may be used instead of Methods 1 and 2. The "Enclosing Rectangle" and "Aggregate Notional Area" methods are included in the BRE report.

Basis for calculating acceptable unprotected area

13.16 The basis of Methods 1 and 2 is set out in Fire Research Technical Paper No 5, 1963. This has been reprinted as part of the BRE report referred to in paragraph 13.15. The aim is to ensure that the building is separated from the boundary by at least half the distance at which the total thermal radiation intensity received from all unprotected areas in the wall would be 12.6 kw/m^2 (in still air), assuming the radiation intensity at each unprotected area is:

Diagram 46 Permitted unprotected areas in small residential buildings

See para 13.19

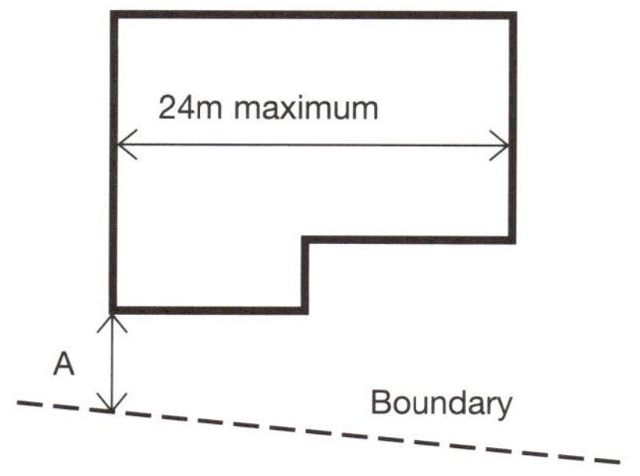

Minimum distance (A) between side of building and relevant boundary (m)	Maximum total area of unprotected areas (m^2)
1	5.6
2	12
3	18
4	24
5	30
6	No limit

a. 84 kw/m^2, if the building is in the Residential, Office or Assembly and Recreation Purpose Groups, or is an open-sided multi-storey car park in Purpose Group 7(b); and

b. 168 kw/m^2, if the building is in the Shop and Commercial, Industrial, Storage or Other non-residential Purpose Groups.

Sprinkler systems

13.17 If a building is fitted throughout with a sprinkler system, it is reasonable to assume that the intensity and extent of a fire will be reduced. The sprinkler system should be in accordance with paragraph 0.16. In these circumstances the boundary distance may be half that for an otherwise similar, but unsprinklered, building, subject to there being a minimum distance of 1m. Alternatively, the amount of unprotected area may be doubled if the boundary distance is maintained.

Note: The presence of sprinklers may be taken into account in a similar way when using the BRE report referred to in paragraph 13.15.

Atrium buildings

13.18 If a building contains one or more atria, the recommendations of clause 28.2 in BS 5588-7:1997 should be followed.

Method 1

13.19 This method applies only to a building intended to be used for block of flats or other residential purposes (not Institutional), which is 1000mm or more from any point on the relevant boundary.

The following rules for determining the maximum unprotected area should be read with Diagram 46.

a. The building should not exceed 3 storeys in height (basements not counted) or be more than 24m in length:

b. Each side of the building will meet the provisions for space separation if:

   i. the distance of the side of the building from the relevant boundary; and

   ii the extent of the unprotected area, are within the limits given in Diagram 46.

   Note: In calculating the maximum unprotected area, any areas falling within the limits shown in Diagram 44 and referred to in paragraph 13.10, can be disregarded.

c. Any parts of the side of the building in excess of the maximum unprotected area should be fire-resisting.

Method 2

13.20 This method applies to a building or compartment intended for any use and which is not less than 1000mm from any point on the relevant boundary.

The following rules for determining the maximum unprotected area should be read with Table 15.

a. The building or compartment should not exceed 10m in height except for an open-sided car park in Purpose Group 7(b) (see paragraph 11.3).

   Note: For any building or compartment more than 10m in height, the methods set out in the BRE report *External fire spread: Building separation and boundary distances* can be applied.

b. Each side of the building will meet the provisions for space separation if either:

   i. the distance of the side of the building from the relevant boundary; and

   ii. the extent of unprotected area, are within the appropriate limits given in Table 15.

   Note: In calculating the maximum unprotected area, any areas shown in Diagram 44 and referred to in paragraph 13.10, can be disregarded.

c. any parts of the side of the building in excess of the maximum unprotected area should be fire-resisting.

Table 15 Permitted unprotected areas in small buildings or compartments

Minimum distance between side of building and relevant boundary (m)		Maximum total percentage of unprotected area %
Purpose groups		
Residential, office, assembly and recreation	Shop and commercial industrial, storage and other non-residential	
(1)	(2)	(3)
n.a.	1	4
1	2	8
2.5	5	20
5	10	40
7.5	15	60
10	20	80
12.5	25	100

Notes:

n.a. = not applicable

a. Intermediate values may be obtained by interpolation.

b. For buildings which are fitted throughout with an automatic sprinkler system, see para 13.17.

c. In the case of open-sided car parks in Purpose Group 7(b), the distances set out in column (1) may be used instead of those in column (2).

d. The total percentage of unprotected area is found by dividing the total unprotected area by the area of a rectangle that encloses all the unprotected areas and multiplying the result by 100.

Arson prevention

13.21 An understanding of the problem and potential occurrence of arson is vital for the building designer, who should be aware that, in industry and commerce, more than 50% of fires are started deliberately, some of them by members of the workforce.

The FPA's book *The prevention and control of arson* deals fully with the subject and includes a chapter entitled 'Designing against arson'. It is a major topic to be considered in the risk assessment, particularly if the location of the building is an area where arson is common.

Those involved in the design and construction of buildings need to be aware of the gravity of the problem and to bear in mind at least the following so that a new building will be better protected against the possibility of an arson attack.

- Potential arsonists may be kept at bay by properly organised physical and electronic security measures, including perimeter fencing, secure doors and windows, an intruder alarm system and even monitored CCTV.
- Adequate security is of particular importance when a building is located in a rural area, when the fire brigade may take longer to respond than in town.
- Some buildings, such as warehouses or superstores, may have accumulated quantities of delivery pallets (plastic and wooden) on the premises. These are favourite targets of the arsonist and provision should be made for secure storage areas outdoors. These should be at least 10m from the outer walls of the warehouse, plant or adjacent premises' buildings and at least 2m from the perimeter fence. If this cannot be achieved, the external walls should have fire resistance as shown in Appendix A, Table A9.
- Waste materials are attractive to the arsonist and a code of practice for dealing with commercial and industrial waste is included in the FPA's book *Fire safety and waste materials*.
- Vegetation and undergrowth in the proximity of the warehouse or alongside the perimeter fence should be pruned regularly so as not to provide a potential source of fuel or a place of concealment for arsonists or to help intruders scale fences.

Section 14: Roof coverings

Introduction

14.1 The provisions in this section limit the use, near a boundary, of roof coverings which will not give adequate protection against the spread of fire over them. The term roof covering is used to describe constructions which may consist of one or more layers of material, but does not refer to the roof structure as a whole. The provisions in this Section are principally concerned with the performance of roofs when exposed to fire from the outside.

14.2 The circumstances when a roof is subject to the provisions in Section 13 for space separation are explained in paragraph 13.1.

Other controls on roofs

14.3 There are provisions concerning the fire properties of roofs in three other Sections of this document. In the guidance to B1 (paragraph 5.3) there are provisions for roofs that are part of a means of escape. In the guidance to B2 (paragraph 6.12) there are provisions for the internal surfaces of rooflights as part of the internal lining of a room or circulation space. In the guidance to B3 there are provisions in Section 7 for roofs which are used as a floor and in Section 8 for roofs that pass over the top of a compartment wall.

Classification of performance

14.4 The performance of roof coverings is designated by reference to the test methods specified in BS 476-3: 2004 or determined in accordance with BS EN 13501-5:2005, as described in Appendix A. The notional performance of some common roof coverings is given in Table A5 of Appendix A.

Rooflights are controlled on a similar basis and plastic rooflights described in paragraph 14.6 and 14.7 may also be used.

Separation distances

14.5 The separation distance is the minimum distance from the roof (or part of the roof) to the relevant boundary, which may be a notional boundary.

Table 16 sets out separation distances according to the type of roof covering and the size and use of the building. There are no restrictions on the use of roof coverings designated AA, AB or AC (National class) or B_{ROOF}(t4) (European class) classification. In addition, roof covering products (and/or materials) as defined in Commission Decision 2000/553/EC of 6th September 2000 implementing Council Directive 89/106/EEC as regards the external fire performance of roof coverings can be considered to fulfil all of the requirements for performance characteristic "external fire performance" without the need for testing **provided that any national provisions on the design and execution of works are fulfilled**. That is, the roof covering products (and/or materials) defined in this Commission Decision can be used without restriction.

Diagram 47 **Limitations on spacing and size of plastic rooflights having a Class 3 (National class) or Class D-s3, d2 (European class) or TP(b) lower surface**

See para 14.6

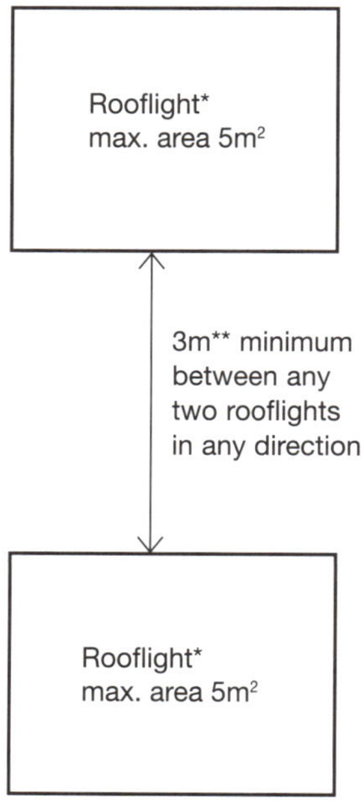

* Or group of rooflights amounting to no more than $5m^2$

** Class 3 rooflights to rooms in industrial and other non-residential purpose groups may be spaced 1800mm apart provided the rooflights are evenly distributed and do not exceed 20% of the area of the room

Notes:

1. There are restrictions on the use of plastic rooflights in the guidance to B2.
2. Surrounding roof covering to be a material of limited combustibility for at least 3m distance.
3. Where Diagram 30a or b applies, rooflights should be at least 1500mm from the compartment wall.

Plastic rooflights

14.6 Table 17 sets out the limitations on the use of plastic rooflights which have at least a Class 3 (National class) or Class D-s3, d2 (European class) lower surface and Table 18 sets out the limitations on the use of thermoplastic materials with a TP(a) rigid or TP(b) classification (see also Diagram 47). The method of classifying thermoplastic materials is given in Appendix A.

14.7 When used in rooflights, a rigid thermoplastic sheet product made from polycarbonate or from unplasticised PVC, which achieves a Class 1 (National class) rating for surface spread of flame when tested to BS 476-7:1971 (or 1987 or 1997) *Surface spread of flame tests for* materials, or Class C-s3, d2 (European class) can be regarded as having an AA (National class) designation or B_{ROOF}(t4) (European class) classification, other than for the purposes of Diagram 30.

Unwired glass in rooflights

14.8 When used in rooflights, unwired glass at least 4mm thick can be regarded as having an AA designation (National class) designation or B_{ROOF}(t4) (European class) classification.

Thatch and wood shingles

14.9 Thatch and wood shingles should be regarded as having an AD/BD/CD designation or E_{ROOF}(t4) (European class) classification in Table 16 if performance under BS 476-3: 1958 2004 or EN 1187:XXX (test 4) respectively cannot be established.

Table 16 Limitations on roof coverings* (see also Table 16a)

Designation† of covering of roof or part of roof		Minimum distance from any point on relevant boundary			
National Class	European Class	Less than 6m	At least 6m	At least 12m	At least 20m
AA, AB or AC	B_{ROOF}(t4)	●	●	●	●
BA, BB or BC	C_{ROOF}(t4)	○	●	●	●
CA, CB or CC	D_{ROOF}(t4)	○	●(1) (2)	●(1)	●
AD, BD or CD	E_{ROOF}(t4)	○	●(1) (2)	●(1)	●(1)
DA, DB, DC or DD	F_{ROOF}(t4)	○	○	○	●(1) (2)

Notes:

* See paragraph 14.8 for limitations on glass; paragraph 14.9 for limitations on thatch and wood shingles; and paragraphs 14.6 and 14.7 and Tables 18 and 19 for limitations on plastic rooflights.

† The designation of external roof surfaces is explained in Appendix A. (See Table A5, for notional designations of roof coverings.)

Separation distances do not apply to the boundary between roofs of a pair of semi-detached houses (see 14.5) and to enclosed/covered walkways. However, see Diagram 30 if the roof passes over the top of a compartment wall. Polycarbonate and PVC rooflights which achieve a Class 1 rating by test, see paragraph 15.7, may be regarded as having an AA designation.

Openable polycarbonate and PVC rooflights which achieve a Class 1 (National class) or Class C-s3, d2 (European class) rating by test, see paragraph 10.7, may be regarded as having an AA (National class) designation or B_{ROOF}(t4) (European class) classification.

● Acceptable.

○ Not acceptable.

1. Not acceptable on any of the following buildings:
   a. Houses in terraces of three or more houses.
   b. Industrial, storage or other non-residential Purpose Group buildings of any size.
   c. Any other buildings with a cubic capacity of more than 1500m³.
2. Acceptable on buildings not listed in Note 1, if part of the roof is no more than 3m² in area and is at least 1500mm from any similar part, with the roof between the parts covered with a material of limited combustibility.

Table 16a Property protection – limitations on roof coverings

Designation of covering of roof or part of roof		Minimum distance from any point on relevant boundary			
National Class	European Class	Less than 6m or high financial exposure	At least 6m	At least 12m	At least 20m
AA	N/A	●	●	●	●
AB	N/A	○	●	●	●
AC	B_{ROOF}(t4)	○	○	●	●
BA, BB or BC	C_{ROOF}(t4)	○	○	○	○
CA, CB or CC	D_{ROOF}(t4)	○	○	○	○
AD, BD or CD	E_{ROOF}(t4)	○	○	○	○
DA, DB, DC or DD	F_{ROOF}(t4)	○	○	○	○

Notes:

The test and classification of BS 476-3:2004 is referenced in LPS 1505 thus maintaining the penetration and flame spread classification. The latter classification is not included in the European class.

● Acceptable.

○ Not acceptable.

The risk from hot work needs consideration when specifying materials used in the construction of roofs (roof coverings, insulants and substrates). Maintenance work involving hot work processes such as welding, plumbing and paint stripping often results in fire and, as such, can lead to damage to property and business interruption. Similar risk of fire can occur during the construction process.

In addition, the risks imposed by storage of combustible materials between buildings (see Diagram 42a) has been considered in Table 16a.

Table 17 Class 3 (National class) or Class D-s3, d2 (European class) plastic rooflights: limitations on use and boundary distance

Minimum classification on lower surface (1)	Space which rooflight can serve	Minimum distance from any point on relevant boundary to rooflight with an external designation† of:	
		AD BD CD (National class) or E_{ROOF}(t4) (European class) CA CB CC or D_{ROOF}(t4) (European class)	DA DB DC DD (National class) or F_{ROOF}(t4) (European class)
Class 3	a. Balcony, verandah, carport, covered way or loading bay, which has at least one longer side wholly or permanently open b. Detached swimming pool c. Conservatory, garage or outbuilding, with a maximum floor area of 40m²	6m	20m
	d. Circulation space (2) (except a protected stairway) e. Room (2)	6m (3)	20m (3)

Notes:

† The designation of external roof surfaces is explained in Appendix A.

None of the above designations are suitable for protected stairways – see paragraph 6.12.

Polycarbonate and PVC rooflights which achieve a Class 1 (National class) or Class C-s3, d2 (European class) rating by test, see paragraph 14.7, may be regarded as having an AA designation or B_{ROOF}(t4) (European class) classification.

Where Diagram 30a or b applies, rooflights should be at least 1.5m from the compartment wall (2.5m is minimum for property protection).

Products may have upper & lower surfaces with different properties if they have double skins or are laminates of different materials. In which case the more onerous distance applies.

1. See also the guidance to B2.
2. Single skin rooflight only, in the case of non-thermoplastic material.
3. The rooflight should also meet the provisions of Diagram 47.

Table 18 TP(a) and TP(b) plastic rooflights: limitations on use and boundary distance

Minimum classification on lower surface [1]	Space which rooflight can serve	Minimum distance from any point on relevant boundary to rooflight with an external surface classification [1] of:	
		TP(a)	TP(b)
1. TP(a) rigid	Any space except a protected stairway or area of special risk or any sprinkler-protected area	6m [2]	Not applicable
2. TP(b)	a. Balcony, verandah, carport, covered way or loading bay, which has at least one longer side wholly or permanently open b. Detached swimming pool c. Conservatory, garage or outbuilding, with a maximum floor area of 40m^2	Not applicable	6m
	d. Circulation space [3] (except a protected stairway) e. Room [3]	Not applicable	6m [4]

Notes:

None of the above designations are suitable for protected stairways – see paragraph 6.12.

Polycarbonate and PVC rooflights which achieve a Class 1 rating by test, see paragraph 14.7, may be regarded as having an AA designation.

Where Diagram 30a or b applies, rooflights should be at least 1.5m from the compartment wall (2.5m is minimum for property protection).

Products may have upper and lower surfaces with different properties if they have double skins or are laminates of different materials; in which case the more onerous distance applies.

1. See also the guidance to B2.
2. No limit in the case of any space described in 2a, b and c.
3. Single skin rooflight only, in the case of non-thermoplastic material.
4. The rooflight should also meet the provisions of Diagram 47.

For sprinklered risks, no thermoplastic rooflights TP(a) rigid or TP (b) rooflights should be specified. This is particularly important with ESFR (early suppression fast response) sprinklers.

There are two issues with sprinklers and plastic rooflights:

1. *If plastic rooflights lose their strength and rigidity, they may become detached and drape over a sprinkler below the rooflight opening;*
2. *If a rooflight falls out it may create similar problems to a smoke vent, i.e. the hot gases may be vented and delay the operation of the sprinkler(s). This is particularly critical for ESFR sprinklers where any delay in sprinkler operation could jeopardize the effectiveness of the protection, hence the 300°C requirement (see* LPC Rules for automatic sprinkler installations*). Consequently any rooflights with low temperature characteristics would be unacceptable for ESFR.*

The Requirement

This Approved Document deals with the following Requirement from Part B of Schedule 1 to the Building Regulations 2000 (as amended).

Requirement	*Limits on application*
Access and facilities for the fire service **B5.** (1) The building shall be designed and constructed so as to provide reasonable facilities to assist firefighters in the protection of life. (2) Reasonable provision shall be made within the site of the building to enable fire appliances to gain access to the building.	

Guidance

Performance

In the Secretary of State's view the Requirements of B5 will be met:

a. if there is sufficient means of external access to enable fire appliances to be brought near to the building for effective use;

b. if there is sufficient means of access into and within, the building for firefighting personnel to effect search and rescue and fight fire;

c. if the building is provided with sufficient internal fire mains and other facilities to assist firefighters in their tasks; and

d. if the building is provided with adequate means for venting heat and smoke from a fire in a basement.

These access arrangements and facilities are only required in the interests of the health and safety of people in and around the building. The extent to which they are required will depend on the use and size of the building in so far as it affects the health and safety of those people.

Introduction

B5.i The guidance given here covers the selection and design of facilities for the purpose of protecting life by assisting the fire and rescue service. To assist the fire and rescue service some or all of the following facilities may be necessary, depending mainly on the size of the building:

a. vehicle access for fire appliances;

b. access for firefighting personnel;

c. the provision of fire mains within the building;

d. venting for heat and smoke from basement areas;

e. the provision of adequate water supplies.

If it is proposed to deviate from the general guidance in sections 15 to 18 then it would be advisable to seek advice from the relevant Fire and Rescue Service at the earliest opportunity, even where there is no statutory duty to consult.

Facilities appropriate to a specific building

B5.ii The main factor determining the facilities needed to assist the fire and rescue service is the size of the building. Generally speaking firefighting is carried out within the building.

a. In deep basements and tall buildings (see paragraph 17.2) firefighters will invariably work inside. They need special access facilities (see Section 17), equipped with fire mains (see Section 15). Fire appliances will need access to entry points near the fire mains (see Section 16).

b. In other buildings, the combination of personnel access facilities offered by the normal means of escape and the ability to work from ladders and appliances on the perimeter, will generally be adequate without special internal arrangements. Vehicle access may be needed to some or all of the perimeter, depending on the size of the building (see Section 15).

   Note: Where an alternative approach outside the scope of this Approved Document has been used to justify the means of escape it may be necessary to consider additional provisions for firefighting access.

c. For small buildings, it is usually only necessary to ensure that the building is suffiently close to a point accessible to fire and rescue service vehicles (see paragraph 16.2).

d. In taller blocks of flats, fire and rescue service personnel access facilities are needed within the building, although the high degree of compartmentation means that some simplification is possible compared to other tall buildings (see paragraph 17.12);

e. Products of combustion from basement fires tend to escape via stairways, making access difficult for fire and rescue service personnel. The problem can be reduced by providing vents (see Section 17). Venting can improve visibility and reduce temperatures, making search, rescue and firefighting less difficult.

Insulating core panels

B5.iii Guidance on the fire behaviour of insulating core panels used for internal structures is given in Appendix F. **Additional information regarding insurers' concerns is also included in this document.**

Section 15: Fire mains and hydrants

Introduction

15.1 Fire mains are installed in a building and equipped with valves etc so that the fire and rescue service may connect hoses for water to fight fires inside the building.

Fire mains may be of the 'dry' type which are normally empty and are supplied through hose from a fire and rescue service pumping appliance. Alternatively, they may be of the 'wet' type where they are kept full of water and supplied from tanks and pumps in the building. There should be a facility to allow a wet system to be replenished from a pumping appliance in an emergency.

Provision of fire mains

15.2 Buildings with firefighting shafts should be provided with fire mains in those shafts and, where necessary, in protected escape stairs. The criteria for the provision of firefighting shafts and fire mains in such buildings are given in Section 17.

15.3 Fire mains may also be provided in other buildings where vehicle access is not provided in accordance with Table 19 (see paragraphs 16.6 & 16.7) or paragraphs 16.2 or 16.3.

Number and location of fire mains

15.4 In buildings provided with fire mains for the purposes of paragraph 15.3, outlets from fire mains should be located to meet the hose criterion set out in paragraph 17.8. This does not imply that these stairs need to be designed as firefighting shafts.

Design and construction of fire mains

15.5 The outlets from fire mains should be located within the protected enclosure of a stairway or a protected lobby where one is provided (see Diagram 52).

15.6 Guidance on other aspects of the design and construction of fire mains, not included in the provisions of this Approved Document, should be obtained from BS 9990:2006.

Note: Wet fire mains should be provided in buildings with a floor at more than 50m above fire and rescue service vehicle access level. In lower buildings where fire mains are provided, either wet or dry mains are suitable.

Provision of private hydrants

15.7 Where a building, which has a compartment of 280m^2 or more in area, is being erected more than 100m from an existing fire-hydrant additional hydrants should be provided as follows;

a. **Buildings provided with fire mains –** hydrants should be provided within 90m of dry fire main inlets.

b. **Buildings not provided with fire mains –** hydrants should be provided within 90m of an entry point to the building and not more than 90m apart.

Each fire hydrant should be clearly indicated by a plate, affixed nearby in a conspicuous position, in accordance with BS 3251:1976.

15.8 Where no piped water supply is available, or there is insufficient pressure and flow in the water main, or an alternative arrangement is proposed, the alternative source of supply should be provided in accordance with the following recommendations:

a. a charged static water tank of at least 45,000 litre capacity; or

b. a spring, river, canal or pond capable of providing or storing at least 45,000 litres of water at all times of the year, to which access, space and a hard standing are available for a pumping appliance; or

c. any other means of providing a water supply for firefighting operations considered appropriate by the fire and rescue authority.

Section 16: Vehicle access

Introduction

16.1 For the purposes of this Approved Document vehicle access to the exterior of a building is needed to enable high reach appliances, such as turntable ladders and hydraulic platforms, to be used and to enable pumping appliances to supply water and equipment for firefighting, search and rescue activities.

Access requirements increase with building size and height.

Fire mains (see Section 15) enable firefighters within the building to connect their hoses to a water supply. In buildings fitted with fire mains, pumping appliances need access to the perimeter at points near the mains, where firefighters can enter the building and where in the case of dry mains, a hose connection will be made from the appliance to pump water into the main.

The vehicle access requirements described in Table 19 for buildings without fire mains, do not apply to blocks of flats, because access is required to each individual dwelling (see paragraph 16.3), or to buildings with fire mains.

Vehicle access routes and hard-standings should meet the criteria described in paragraphs 16.7 to 16.10 where they are to be used by fire and rescue service vehicles.

Note: Requirements cannot be made under the Building Regulations for work to be done outside the site of the works shown on the deposited plans, building notice or initial notice.

In this connection it may not always be reasonable to upgrade an existing route across a site to a small building. The options in such a case, from doing no work to upgrading certain features of the route e.g. a sharp bend, should be considered by the Building Control Body in consultation with the Fire and Rescue Service.

Buildings not fitted with fire mains

16.2 There should be vehicle access for a pump appliance to small buildings (those of up to 2000m^2 with a top storey up to 11m above ground level) to either:

a. 15% of the perimeter; or

b. within 45m of every point on the projected plan area (or 'footprint', see Diagram 48) of the building; whichever is the less onerous.

16.3 There should be vehicle access for a pump appliance to blocks of flats to within 45m of all points within each dwelling.

Note 1: If the provisions in paragraph 16.2 or 16.3 cannot be met, a fire main should be provided in accordance with paragraph 15.3 and vehicle access should meet paragraph 16.6.

Table 19 **Fire and rescue service vehicle access to buildings (excluding blocks of flats) not fitted with fire mains**

Total floor area [(1)] of building m²	Height of floor of top storey above ground [(2)]	Provide vehicle access [(3)] [(4)] to:	Type of appliance
Up to 2000	Up to 11 Over 11	See paragraph 16.2 15% of perimeter [(5)]	Pump High reach
2000–8000	Up to 11 Over 11	15% of perimeter [(5)] 50% of perimeter [(5)]	Pump High reach
8000–16,000	Up to 11 Over 11	50% of perimeter [(5)] 50% of perimeter [(5)]	Pump High reach
16,000–24,000	Up to 11 Over 11	75% of perimeter [(5)] 75% of perimeter [(5)]	Pump High reach
Over 24,000	Up to 11 Over 11	100% of perimeter [(5)] 100% of perimeter [(5)]	Pump High reach

Notes:

1. The total floor area is the aggregate of all floors in the building (excluding basements).
2. In the case of Purpose Group 7(a) (storage) buildings, height should be measured to mean roof level, see Methods of measurement in Appendix C.
3. An access door is required to each such elevation (see paragraph 16.5).
4. See paragraph 16.8 for meaning of access.
5. Perimeter is decribed in Diagram 48.

Note: For single-storey commercial and industrial buildings not fitted with fire mains, the requirements of Table 19a should be followed.

Table 19a Fire and rescue service vehicle access to single-storey commercial and industrial buildings not fitted with fire mains

Total floor area of building (m^2)	Height of building (m)	Provide vehicle access to: (1) (2)	Type of appliance
Up to 2000	Up to 11 Over 11	25% of perimeter 25% of perimeter	Pump High reach
2000–8000	Up to 11 Over 11	50% of perimeter 50% of perimeter	Pump High reach
8000–16,000	Up to 11 Over 11	75% of perimeter 75% of perimeter	Pump High reach
Over 16,000	Up to 11 Over 11	100% of perimeter 100% of perimeter	Pump High reach

Notes:

1. An access door is required to each such elevation (see paragraph 16.5).
2. See paragraph 16.8 for meaning of access.

Diagram 48 **Example of building footprint and perimeter**

See para 16.2 and Table 19

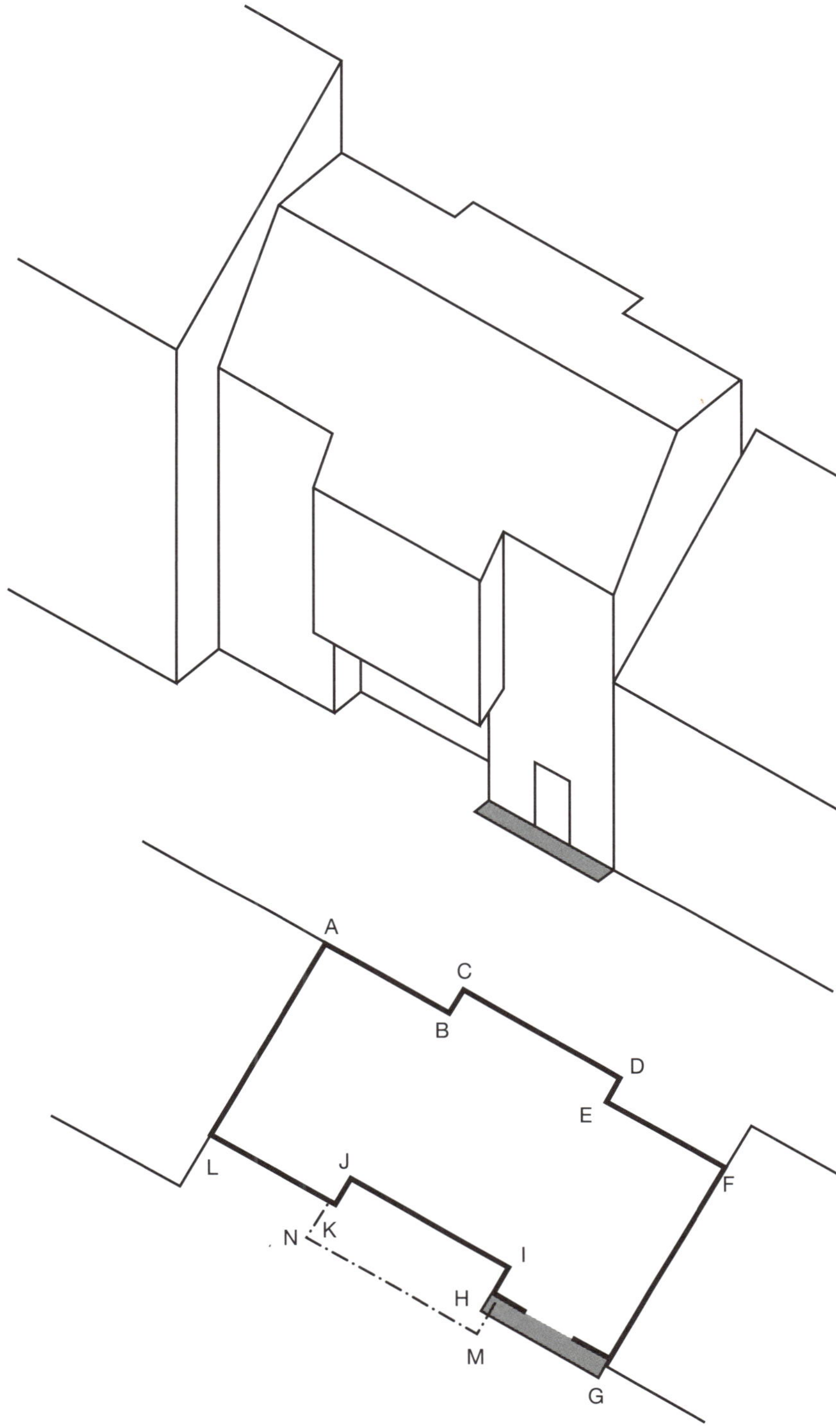

Plan of building AFGL where AL and FG are walls in common with other buildings.
The footprint of the building is the maximum aggregate plan perimeter found by the vertical projection of any overhanging storey onto a ground storey (i.e. ABCDEFGHMNKL).

The perimeter of the building for the purposes of Table 20 is the sum of the lengths of the two external walls, taking account of the footprint i.e. (A to B to C to D to E to F) + (G to H to M to N to K to L).

If the dimensions of the building are such that Table 20 requires vehicle access, the shaded area illustrates one possible example of 15% of the perimeter. **Note:** There should be a door into the building in this length (see paragraph 16.5).

If the building does not have walls in common with other buildings, the lengths AL and FG would be included in the perimeter.

Diagram 49 Relationship between building and hardstanding/access roads for high reach fire appliances

See para 16.9

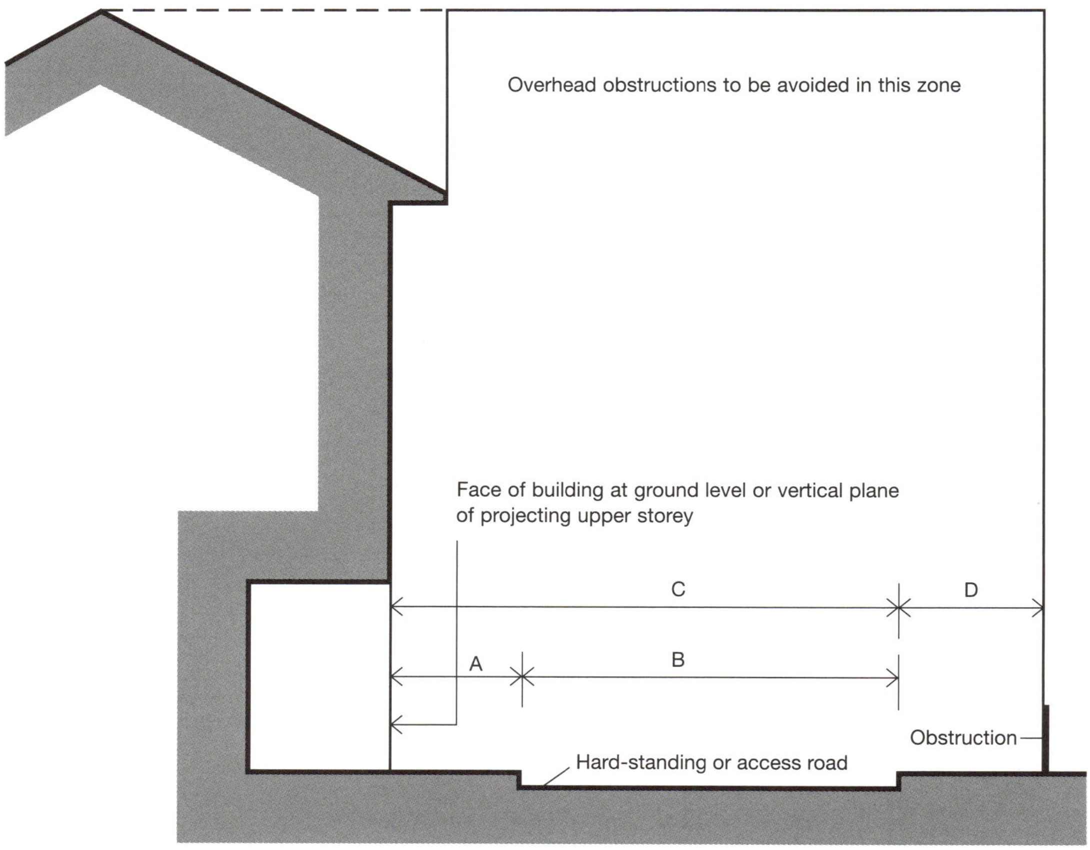

	Type of appliance	
	Turntable ladder dimension (m)	Hydraulic platform dimension (m)
A. Maximum distance of near edge of hard-standing from building	4.9	2.0
B. Minimum width of hardstanding	5.0	5.5
C. Minimum distance of further edge of hard-standing from building	10.0	7.5
D. Minimum width of unobstructed space (for swing of appliance platform)	NA	2.2

Notes:

1 Hard-standing for high reach appliances should be as level as possible and should not exceed a gradient of 1 in 12.

2 Fire appliances are not standardised. Some fire services have appliances with a greater weight or different size. In consultation with the Fire and Rescue Service, the Building Control Body should adopt the relevant dimensions and ground loading capacity.

16.4 Vehicle access to buildings that do not have fire mains (other than buildings described in paragraph 16.2) should be provided in accordance with Table 19.

16.5 Every elevation to which vehicle access is provided in accordance with paragraph 16.2 or Table 19 should have a suitable door(s), not less than 750mm wide, giving access to the interior of the building.

Door(s) should be provided such that there is no more than 60m between each door and/or the end of that elevation (e.g. a 150m elevation would need at least 2 doors).

Table 20 **Typical fire and rescue service vehicle access route specification**

Appliance type	Minimum width of road between kerbs (m)	Minimum width of gateways (m)	Minimum turning circle between kerbs (m)	Minimum turning circle between walls (m)	Minimum clearance height (m)	Minimum carrying capacity (tonnes)
Pump	3.7	3.1	16.8	19.2	3.7	12.5
High reach	3.7	3.1	26.0	29.0	4.0	17.0

Notes:

1. Fire appliances are not standardised. Some fire services have appliances of greater weight or different size. In consultation with the Fire and Rescue Service, the Building Control Body may adopt other dimensions in such circumstances.
2. Because the weight of high reach appliances is distributed over a number of axles, it is considered that their infrequent use of a carriageway or route designed to 12.5 tonnes should not cause damage. It would therefore be reasonable to design the roadbase to 12.5 tonnes, although structures such as bridges should have the full 17 tonnes capacity.

Buildings fitted with fire mains

16.6 In the case of a building fitted with dry fire mains there should be access for a pumping appliance to within 18m of each fire main inlet connection point, typically on the face of the building. The inlet should be visible from the appliance.

16.7 In the case of a building fitted with wet mains the pumping appliance access should be to within 18m and within sight of, a suitable entrance giving access to the main and in sight of the inlet for the emergency replenishment of the suction tank for the main.

Note: Where fire mains are provided in buildings for which Sections 15 and 17 make no provision, vehicle access may be to paragraph 16.6 or 16.7 rather than Table 19.

Design of access routes and hard-standings

16.8 A vehicle access route may be a road or other route which, including any inspection covers and the like, meets the standards in Table 20 and the following paragraphs.

16.9 Where access is provided to an elevation in accordance with Table 19 for:

a. buildings up to 11m in height (excluding buildings covered by paragraph 16.2(b)), there should be access for a pump appliance adjacent to the building for the percentage of the total perimeter specified;

b. buildings over 11m in height, the access routes should meet the guidance in Diagram 49.

16.10 Where access is provided to an elevation for high reach appliances in accordance with Table 19, overhead obstructions such as cables and branches that would interfere with the setting of ladders etc, should be avoided in the zone shown in Diagram 49.

16.11 Turning facilities should be provided in any dead-end access route that is more than 20m long (see Diagram 50). This can be by a hammerhead or turning circle, designed on the basis of Table 20.

Diagram 50 **Turning facilities**

See para 16.11

Fire and rescue service vehicles should not have to reverse more than 20m from the end of an access road

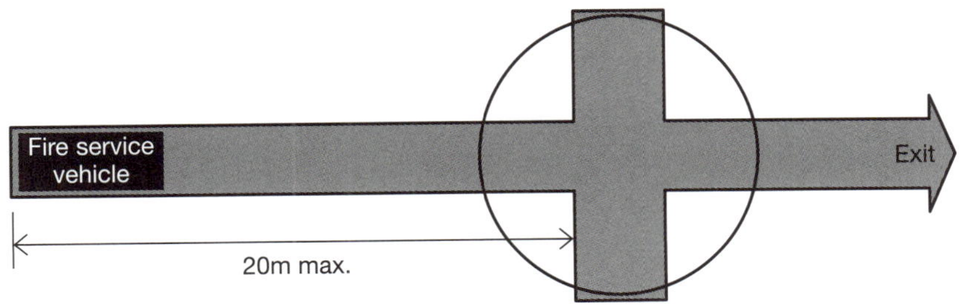

Turning circle, hammerhead or other point at which vehicle can turn

Section 17: Access to buildings for firefighting personnel

Introduction

17.1 In low-rise buildings without deep basements fire and rescue service personnel access requirements will be met by a combination of the normal means of escape and the measures for vehicle access in Section 16, which facilitate ladder access to upper storeys. In other buildings, the problems of reaching the fire and working inside near the fire, necessitate the provision of additional facilities to avoid delay and to provide a sufficiently secure operating base to allow effective action to be taken.

These additional facilities include firefighting lifts, firefighting stairs and firefighting lobbies, which are combined in a protected shaft known as the firefighting shaft (Diagram 52).

Guidance on protected shafts in general is given in Section 8.

Note: Because of the high degree of compartmentation in blocks of block of flats, the provisions for the design and construction of firefighting shafts is different to other buildings.

Provision of firefighting shafts

17.2 Buildings with a floor at more than 18m above fire and rescue service vehicle access level, or with a basement at more than 10m below fire and rescue service vehicle access level, should be provided with firefighting shafts containing firefighting lifts (see Diagram 51).

17.3 Buildings in Purpose Groups 4, 5 and 6 with a storey of 900m^2 or more in area, where the floor is at a height of more than 7.5m above fire and rescue service vehicle access level, should be provided with firefighting shaft(s), which need not include firefighting lifts.

17.4 Buildings with two or more basement storeys, each exceeding 900m^2 in area, should be provided with firefighting shaft(s), which need not include firefighting lifts.

17.5 If a firefighting shaft is required to serve a basement it need not also serve the upper floors unless they also qualify because of the height or size of the building. Similarly a shaft serving upper storeys need not serve a basement which is not large or deep enough to qualify in its own right. However, a firefighting stair and any firefighting lift should serve all intermediate storeys between the highest and lowest storeys that they serve.

17.6 Firefighting shafts should serve all floors through which they pass.

From an insurer perspective, the recommendations from Table 4 from BS 5588-5:2004 should be followed:

Type of building	Recommended content of firefighting shaft
Buildings or parts of buildings where the height of the surface of the floor of the topmost storey (excluding any storey consisting entirely of plant rooms) exceeds 18m	Firefighting stair Firefighting lobbies provided with a fire main Firefighting lift installation
Buildings where the depth of the surface of the floor of the lowermost storey exceeds 10m	Firefighting stair Firefighting lobbies provided with a fire main Firefighting lift installation
Buildings with a height of 11m or more, but less than 18m	Escape stair Unvented firefighting lobby provided with a fire main
Buildings intended to be used as shops, factories or for storage of goods or materials where the height of the topmost storey exceeds 7.5m, with the floor area of any storey above the ground storey not less than 900m^2	Firefighting stair Firefighting lobbies provided with a fire main
Buildings where there are two or more basement levels, each with a floor area exceeding 900m^2	Firefighting stair Firefighting lobbies provided with a fire main

Diagram 51 **Provision of firefighting shafts**

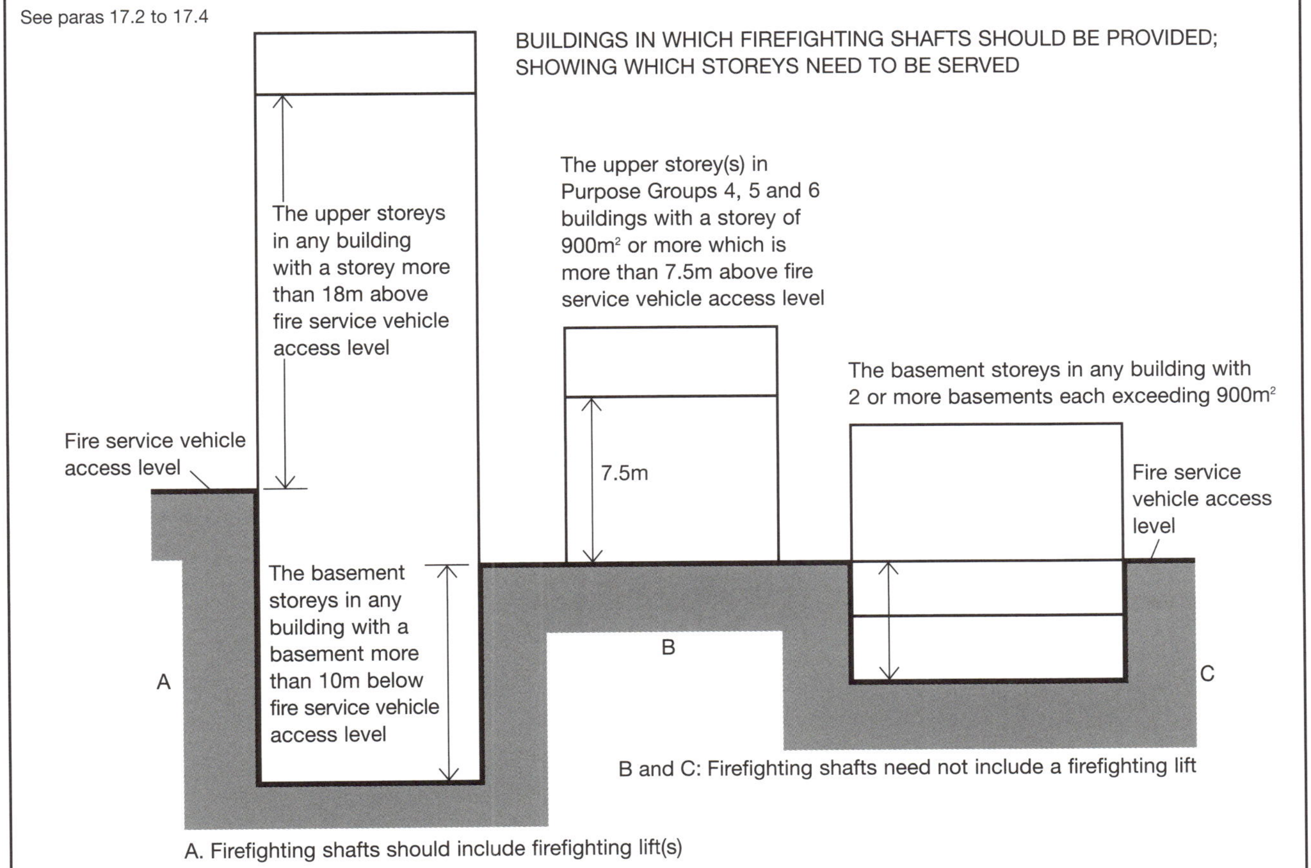

Note: Height excludes any top storey(s) consisting exclusively of plant rooms.

17.7 Shopping complexes should be provided with firefighting shafts in accordance with the recommendations of Section 3 of BS 5588-10:1991.

Number and location of firefighting shafts

17.8 Fire fighting shafts should be located to meet the maximum hose distances set out on paragraph 17.9 or 17.10 and at least two should be provided in buildings with a storey of 900m^2 or more in area, where the floor is at a height of more than 18m above fire and rescue service vehicle access level (or above 7.5m where covered by paragraph 17.3)

17.9 If the building is fitted throughout with an automatic sprinkler system in accordance with paragraph 0.16, then sufficient firefighting shafts should be provided such that every part of every storey, that is more than 18m above fire and rescue service vehicle access level (or above 7.5m where covered by paragraph 17.3), is no more than 60m from a fire main outlet in a firefighting shaft, measured on a route suitable for laying hose.

17.10 If the building is not fitted with sprinklers then every part of every storey that is more than 18m above fire and rescue service vehicle access level (or above 7.5m where covered by paragraph 17.3), should be no more than 45m from a fire main outlet contained in a protected stairway and 60m from a fire main in a firefighting shaft, measured on a route suitable for laying hose.

Note: In order to meet the 45m hose criterion it may be necessary to provide additional fire mains in escape stairs. This does not imply that these stairs need to be designed as firefighting shafts.

Design and construction of firefighting shafts

17.11 Except in blocks of flats (see paragraph 17.14), every firefighting stair and firefighting lift should be approached from the accommodation, through a firefighting lobby.

17.12 All firefighting shafts should be equipped with fire mains having outlet connections and valves at every storey.

17.13 A firefighting lift installation includes the lift car itself, the lift well and the lift machinery space, together with the lift control system and the lift communications system. The shaft should be constructed generally in accordance with clauses 7 and 8 of BS 5588-5:2004. Firefighting lift installations should conform to BS EN 81-72:2003 and to BS EN 81-1:1998 or BS EN 81-2:1998 as appropriate for the particular type of lift.

Diagram 52 **Components of a firefighting shaft**

See para 17.1

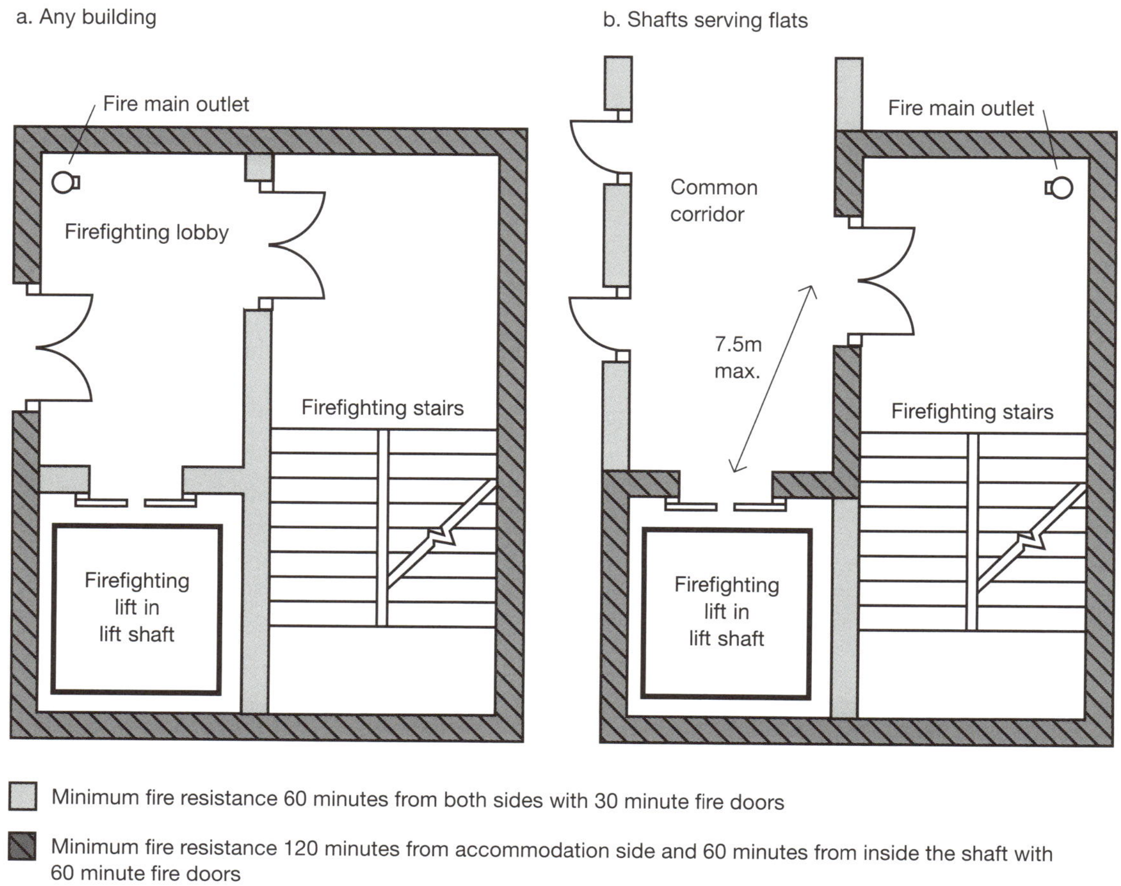

Notes:

1. Outlets from a fire main should be located in the firefighting lobby or, in the case of a shaft serving flats, in the firefighting stairway (see Diagram b).
2. Smoke control should be provided in accordance with BS 5588-5:2004 or, where the shaft only serves flats, the provisions for smoke control given in paragraph 2.25 may be followed instead.
3. A firefighting lift is required if the building has a floor more than 18m above, or more than 10m below, fire service vehicle access level.
4. This Diagram is only to illustrate the basic components and is not meant to represent the only acceptable layout. The shaft should be constructed generally in accordance with clauses 7 and 8 of BS 5588-5:2004.

Variations for block of flats

17.14 Where the design of means of escape in case of fire and compartmentation in blocks of flats has followed the guidance in Sections 3 and 9, the addition of a firefighting lobby between the firefighting stair(s) and the protected corridor or lobby provided for means of escape purposes is not necessary. Similarly, the firefighting lift can open directly into such protected corridor or lobby, but the firefighting lift landing doors should not be more than 7.5m from the door to the firefighting stair.

Rolling shutters in compartment walls

17.15 Rolling shutters should be capable of being opened and closed manually by the fire and rescue service without the use of a ladder.

Section 18: Venting of heat and smoke from basements

Introduction

18.1 The build-up of smoke and heat as a result of a fire can seriously inhibit the ability of the fire and rescue service to carry out rescue and firefighting operations in a basement. The problem can be reduced by providing facilities to make conditions tenable for firefighters.

18.2 Smoke outlets (also referred to as smoke vents) provide a route for heat and smoke to escape to the open air from the basement level(s). They can also be used by the fire and rescue service to let cooler air into the basement(s). (See Diagram 53.)

Provision of smoke outlets

18.3 Where practicable each basement space should have one or more smoke outlets, but it is not always possible to do this where, for example, the plan is deep and the amount of external wall is restricted by adjoining buildings. It is therefore acceptable to vent spaces on the perimeter and allow other spaces to be vented indirectly by opening connecting doors. However if a basement is compartmented, each compartment should have direct access to venting, without having to open doors etc into another compartment.

18.4 Smoke outlets, connected directly to the open air, should be provided from every basement storey, except for any basement storey that has:

a. a floor area of not more than 200m^2 and

b. a floor not more than 3m below the adjacent ground level.

18.5 Strong rooms need not be provided with smoke outlets.

Diagram 53 **Fire-resisting construction for smoke outlet shafts**

See para 18.2

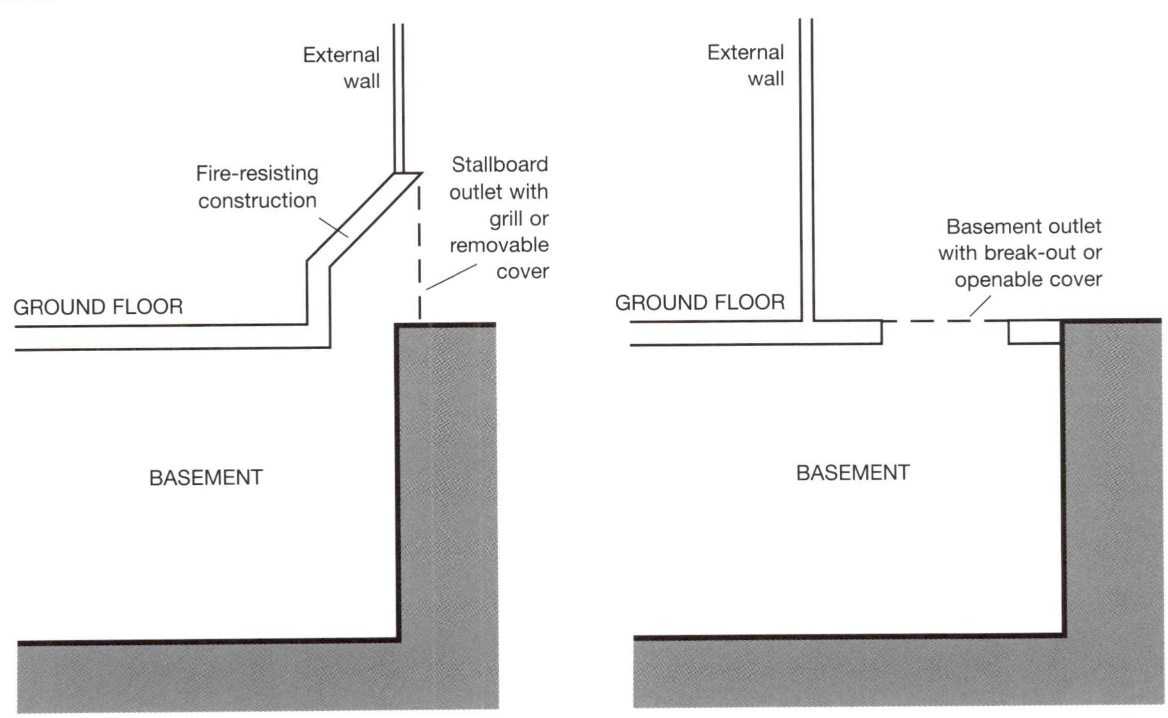

18.6 Where basements have external doors or windows, the compartments containing the rooms with these doors or windows do not need smoke outlets. It is common for basements to be open to the air on one or more elevations. This may be the result of different ground levels on different sides of the building. It is also common in 18th and 19th century terraced housing where an area below street level is excavated at the front and/or rear of the terrace so that the lowest storey has ordinary windows and sometimes an external door.

Natural smoke outlets

18.7 Smoke outlets should be sited at high level, either in the ceiling or in the wall of the space they serve. They should be evenly distributed around the perimeter to discharge in the open air outside the building.

18.8 The combined clear cross-sectional area of all smoke outlets should not be less than 1/40th of the floor area of the storey they serve.

18.9 Separate outlets should be provided from places of special fire hazard.

18.10 If the outlet terminates at a point that is not readily accessible, it should be kept unobstructed and should only be covered with a non-combustible grille or louvre.

18.11 If the outlet terminates in a readily accessible position, it may be covered by a panel, stallboard or pavement light which can be broken out or opened. The position of such covered outlets should be suitably indicated.

18.12 Outlets should not be placed where they would prevent the use of escape routes from the building.

Mechanical smoke extract

18.13 A system of mechanical extraction may be provided as an alternative to natural venting to remove smoke and heat from basements, provided that the basement storey(s) are fitted with a sprinkler system in accordance with paragraph 0.16 (It is not considered necessary in this particular case to install sprinklers on the storeys other than the basement(s) unless they are needed for other reasons.)

Note: Car parks are not normally expected to be fitted with sprinklers (see paragraph 11.2).

18.14 The air extraction system should give at least 10 air changes per hour and should be capable of handling gas temperatures of 300°C for not less than one hour. It should come into operation automatically on activation of the sprinkler system; alternatively activation may be by an automatic fire detection system which conforms to BS 5839-1:2002 (at least L3 standard). For further information on equipment for removing hot smoke refer to BS EN 12101-3:2002.

Construction of outlet ducts or shafts

18.15 Outlet ducts or shafts, including any bulkheads over them (see Diagram 53), should be enclosed in non-combustible construction having not less fire resistance than the element through which they pass.

18.16 Where there are natural smoke outlet shafts from different compartments of the same basement storey, or from different basement storeys, they should be separated from each other by non-combustible construction having not less fire resistance than the storey(s) they serve.

Basement car parks

18.17 The provisions for ventilation of basement car parks in Section 11 may be taken as satisfying the requirements in respect of the need for smoke venting from any basement that is used as a car park.

Appendix A: Performance of materials, products and structures

Introduction

1 Much of the guidance in this document is given in terms of performance in relation to British or European Standards for products or methods of test or design or in terms of European Technical Approvals. In such cases the material, product or structure should:

a. be in accordance with a specification or design which has been shown by test to be capable of meeting that performance; or

   Note: For this purpose, laboratories accredited by the United Kingdom Accreditation Service (UKAS) for conducting the relevant tests would be expected to have the necessary expertise.

b. have been assessed from test evidence against appropriate standards, or by using relevant design guides, as meeting that performance; or

   Note: For this purpose, laboratories accredited by UKAS for conducting the relevant tests and suitably qualified fire safety engineers might be expected to have the necessary expertise.

   For materials/products where European standards or approvals are not yet available and for a transition period after they become available, British standards may continue to be used. Any body notified to the UK Government by the Government of another member state of the European Union as capable of assessing such materials/products against the relevant British Standards, may also be expected to have the necessary expertise. Where European materials/products standards or approvals are available, any body notified to the European Commission as competent to assess such materials or products against the relevant European standards or technical approval can be considered to have the appropriate expertise.

c. where tables of notional performance are included in this document, conform with an appropriate specification given in these tables; or

d. in the case of fire-resisting elements:

   i. conform with an appropriate specification given in Part II of the Building Research Establishment's report *Guidelines for the construction of fire resisting structural elements* (BR 128, BRE 1988); or

   ii. be designed in accordance with a relevant British Standard or Eurocode.

   Note 1: Different forms of construction can present different problems and opportiunities for the provision of structural fire protection. Further information on some specific forms of construction can be found in;

   Timber – BRE 454 *Multi-storey timber frame buildings – a design guide* 2003 (ISBN: 1 86081 605 3)

   Steel – SCI P197 *Designing for structural fire safety: A handbook for architects and engineers* 1999 (ISBN: 1 85942 074 5)

   Note 2: Any test evidence used to substantiate the fire resistance rating of a construction should be carefully checked to ensure that it demonstrates compliance that is adequate and applicable to the intended use. Small differences in detail (such as fixing method, joints, dimensions and the introduction of insulation materials etc.) may significantly affect the rating.

2 Building Regulations deal with fire safety in buildings as a whole. Thus they are aimed at limiting fire hazard. Insurers may require higher levels of performance to protect the property and reduce interruption to the business, in both cases taking into account the potential financial exposure.

The aim of standard fire tests is to measure or assess the response of a material, product, structure or system to one or more aspects of fire behaviour. Standard fire tests cannot normally measure fire hazard. They form only one of a number of factors that need to be taken into account. Other factors are set out in this publication.

Fire resistance

3 Factors having a bearing on fire resistance, that are considered in this document, are:

a. fire severity;

b. building height;

c. building occupancy; and

d. potential to protect the assets of the business.

4 The standards of fire resistance given are based on assumptions about the severity of fires and the consequences should an element fail. Fire severity is estimated in very broad terms from the use of the building (its purpose group), on the assumption that the building contents (which constitute the fire load) are similar for buildings in the same use.

A number of factors affect the standard of fire resistance specified. These are:

a. the amount of combustible material per unit of floor area in various types of building (the fire load density);

b. the height of the top floor above ground, which affects the ease of escape and of firefighting operations and the consequences should large scale collapse occur;

c. occupancy type, which reflects the ease with which the building can be evacuated quickly;

d. whether there are basements, because the lack of an external wall through which to vent heat and smoke may increase heat build-up and thus affect the duration of a fire, as well as complicating firefighting;

e. whether the building is of single storey construction (where escape is direct and structural failure is unlikely to precede evacuation); and

f. insurers will need to consider the risk of fire occurring, the potential area that could be affected by fire and smoke damage, and financial exposures.

Because the use of buildings may change, a precise estimate of fire severity based on the fire load due to a particular use may be misleading. Therefore if a fire engineering approach of this kind is adopted the likelihood that the fire load may change in the future needs to be considered. **For some purpose groups, such as factories and storage buildings, insurance experience is that inception risks (sources of ignition/ potential fire spread) and fire load can vary considerably during the working life of a building and consequently high values of fire load should be assumed.**

5 Performance in terms of the fire resistance to be met by elements of structure, doors and other forms of construction is determined by reference to either:

a. (National tests) BS 476 *Fire tests on building materials and structures,* Parts 20-24: 1987, i.e. Part 20 *Method for determination of the fire resistance of elements of construction (general principles),* Part 21 *Methods for determination of the fire resistance of loadbearing elements of construction,* Part 22 *Methods for determination of the fire resistance of non-loadbearing elements of construction,* Part 23 *Methods for determination of the contribution of components to the fire resistance of a structure* and Part 24 *Method for determination of the fire resistance of ventilation ducts* (or to BS 476-8:1972 in respect of items tested or assessed prior to 1 January 1988); or

b. (European tests) Commission Decision 2000/367/EC of 3rd May 2000 implementing Council Directive 89/106/EEC as regards the classification of the resistance to fire performance of construction products, construction works and parts thereof.

Note: The designation of xxxx is used for the year reference for standards that are not yet published. The latest version of any standard may be used provided that it continues to address the relevant requirements of the Regulations.

All products are classified in accordance with BS EN 13501-2:2003, *Fire classification of construction products and building elements – Classification using data from fire resistance tests (excluding products for use in ventilation systems).*

BS EN 13501-3:2005, *Fire classification of construction products and building elements – Classification using data from fire resistance tests on components of normal building service installations (other than smoke control systems).*

BS EN 13501-4:xxxx, *Fire classification of construction products and building elements – Classification using data from fire resistance tests on smoke control systems.*

The relevant European test methods under BS EN 1364, 1365, 1366 and 1634 are listed in Appendix G.

Table A1 gives the specific requirements for each element in terms of one or more of the following performance criteria:

a. **resistance to collapse** (loadbearing capacity), which applies to loadbearing elements only, denoted R in the European classification of the resistance to fire performance;

b. **resistance to fire penetration** (integrity), denoted E in the European classification of the resistance to fire performance; and

c. **resistance to the transfer of excessive heat** (insulation), denoted I in the European classification of the resistance to fire performance.

Table A2 sets out the minimum periods of fire resistance for elements of structure.

Table A3 sets out criteria appropriate to the suspended ceilings that can be accepted as contributing to the fire resistance of a floor.

Table A4 sets out limitations on the use of uninsulated fire-resisting glazed elements.

These limitations do not apply to the use of insulated fire-resisting glazed elements.

Information on tested elements is frequently given in literature available from manufacturers and trade associations.

Information on tests on fire-resisting elements is also given in such publications as:

Association for Specialist Fire Protection *Fire protection for structural steel in buildings 4th Edition* (ISBN: 1 87040 925 6).

Roofs

6 Performance in terms of the resistance of roofs to external fire exposure is determined by reference to either:

a. (National tests) BS 476-3:2004 *External fire exposure roof tests;* or

b. (European tests) Commission Decision XXXX/YYY/EC amending Decision 2001/671/EC establishing a classification system for the external fire performance of roofs and roof coverings.

Constructions are classified within the National system by 2 letters in the range A to D, with an AA designation being the best. The first letter indicates the time to penetration; the second letter a measure of the spread of flame.

Constructions are classified within the European system as B_{ROOF}(t4), C_{ROOF}(t4), D_{ROOF}(t4), E_{ROOF}(t4) or F_{ROOF}(t4) (with B_{ROOF}(t4) being the highest performance and F_{ROOF}(t4) being the lowest) in accordance with BS EN 13501-5:2005, *Fire classification of construction products and building elements – Classification using test data from external fire exposure to roof tests.*

BS EN 13501-1 refers to four separate tests. The suffix (t4) used above indicates that Test 4 is to be used for the purposes of this Approved Document.

Some roof covering products (and/or materials) can be considered to fulfil all of the requirements for the performance characteristic "external fire performance" without the need for testing, subject to any national provisions on the design and execution of works being fulfilled. These roof covering products are listed in Commission Decision 2000/553/EC of 6th September 2000 implementing Council Directive 89/106/EEC as regards the external fire performance of roof coverings.

In some circumstances roofs, or parts of roofs, may need to be fire-resisting, for example if used as an escape route or if the roof performs the function of a floor. Such circumstances are covered in Sections 2, 4 and 6.

Table A5 gives notional designations of some generic roof coverings.

Reaction to fire

7 Performance in terms of reaction to fire to be met by construction products is determined by Commission Decision 200/147/EC of 8th February 2000 implementing Council Directive 89/106/EEC as regards the classification of the reaction to fire performance of construction products.

Note: The designation of xxxx is used for the year reference for standards that are not yet published. The latest version of any standard may be used provided that it continues to address the relevant requirements of the Regulations.

All products, excluding floorings, are classified as †A1, A2, B, C, D, E or F (with class A1 being the highest performance and F being the lowest) in accordance with BS EN 13501-1:2002, *Fire classification of construction products and building elements,* Part 1 – *Classification using data from reaction to fire tests.*

The relevant European test methods are specified as follows,

BS EN ISO 1182:2002, *Reaction to fire tests for building products – Non-combustibility test.*

BS EN ISO 1716:2002, *Reaction to fire tests for building products – Determination of the gross calorific value.*

BS EN 13823:2002, *Reaction to fire tests for building products – Building products excluding floorings exposed to the thermal attack by a single burning item.*

BS EN ISO 11925-2:2002, *Reaction to fire tests for building Products,* Part 2 – *Ignitability when subjected to direct impingement of a flame.*

BS EN 13238:2001, *Reaction to fire tests for building products – conditioning procedures and general rules for selection of substrates.*

Non-combustible materials

8 Non-combustible materials are defined in Table A6 either as listed products, or in terms of performance:

a. (National classes) when tested to BS 476-4:1970 *Non-combustibility test for materials* or BS 476-11:1982 *Method for assessing the heat emission from building products;* or

b. (European classes) when classified as class A1 in accordance with BS EN 13501-1:2002, *Fire classification of construction products and building elements, Part 1-Classification using data from reaction to fire tests* when tested to BS EN ISO 1182:2002, *Reaction to fire tests for building products – Non-combustibility test* **and** BS EN ISO 1716:2002 *Reaction to fire tests for building products – Determination of the gross calorific value.*

Table A6 identifies non-combustible products and materials and lists circumstances where their use is necessary.

Materials of limited combustibility

9 Materials of limited combustibility are defined in Table A7:

a. (National classes) by reference to the method specified in BS 476: Part 11:1982; or

b. (European classes) in terms of performance when classified as class A2-s3, d2 in accordance with BS EN 13501-1:2002, *Fire classification of construction products and building elements,* Part 1 – *Classification using data from reaction to fire tests* when tested to BS EN ISO 1182:2002, *Reaction to fire tests for building products – Non-combustibility test* or BS EN ISO 1716:2002 *Reaction to fire tests for building products – Determination of the gross calorific value* **and** BS EN 13823:2002, *Reaction to fire tests for building products – Building products excluding floorings exposed to the thermal attack by a single burning item.*

† The classes of reaction to fire performance of A2, B, C, D and E are accompanied by additional classifications related to the production of smoke (s1, s2, s3) and/or flaming droplets/particles (d0, d1, d2).

Table A7 also includes composite products (such as plasterboard) which are considered acceptable and where these are exposed as linings they should also meet any appropriate flame spread rating.

Internal linings

10 Flame spread over wall or ceiling surfaces is controlled by providing for the lining materials or products to meet given performance levels in tests appropriate to the materials or products involved.

11 Under the National classifications, lining systems which can be effectively tested for 'surface spread of flame' are rated for performance by reference to the method specified in BS 476-7:1971 *Surface spread of flame tests for materials,* or 1987 *Method for classification of the surface spread of flame of products,* or 1997

Method of test to determine the classification of the surface spread of flame of products under which materials or products are classified 1, 2, 3 or 4 with Class 1 being the highest.

Under the European classifications, lining systems are classified in accordance with BS EN 13501-1:2002, *Fire classification of construction products and building elements,* Part 1 – *Classification using data from reaction to fire tests.* Materials or products are classified as A1, A2, B, C, D, E or F, with A1 being the highest. When a classification includes "s3, d2", it means that there is no limit set for smoke production and/or flaming droplets/particles.

12 To restrict the use of materials which ignite easily, which have a high rate of heat release and/or which reduce the time to flashover, maximum acceptable 'fire propagation' indices are specified, where the National test methods are being followed. These are determined by reference to the method specified in BS 476-6:1981 or 1989 *Method of test for fire propagation of products.* Index of performance (I) relates to the overall test performance, whereas sub-index (i1) is derived from the first three minutes of test.

13 The highest National product performance classification for lining materials is Class 0. This is achieved if a material or the surface of a composite product is either:

a. composed throughout of materials of limited combustibility; or

b. a Class 1 material which has a fire propagation index (I) of not more than 12 and sub-index (i1) of not more than 6.

Note: Class 0 is not a classification identified in any British Standard test.

Note: Non-combustible or limited combustibility is the highest classification in respect to property protection for lining materials.

14 Composite products defined as materials of limited combustibility (see paragraph 9 above and Table A7) should in addition comply with the test requirement appropriate to any surface rating specified in the guidance on requirements B2, B3 and B4.

15 The notional performance ratings of certain widely used generic materials or products are listed in Table A8 in terms of their performance in the traditional lining tests BS 476 Parts 6 and 7 or in accordance with BS EN 13501-1:2002, *Fire classification of construction products and building elements,* Part 1 – *Classification using data from reaction to fire tests.*

16 Results of tests on proprietary materials are frequently given in literature available from manufacturers and trade associations.

Any reference used to substantiate the surface spread of flame rating of a material or product should be carefully checked to ensure that it is suitable, adequate and applicable to the construction to be used. Small differences in detail, such as thickness, substrate, colour, form, fixings, adhesive etc, may significantly affect the rating.

Thermoplastic materials

17 A thermoplastic material means any synthetic polymeric material which has a softening point below 200°C if tested to BS EN ISO 306:2004 method A120 *Plastics – Thermoplastic materials – Determination of Vicat softening temperature*. Specimens for this test may be fabricated from the original polymer where the thickness of material of the end product is less than 2.5mm.

18 A thermoplastic material in isolation can not be assumed to protect a substrate, when used as a lining to a wall or ceiling. The surface rating of both products must therefore meet the required classification. If however, the thermoplastic material is fully bonded to a non-thermoplastic substrate, then only the surface rating of the composite will need to comply.

19 Concessions are made for thermoplastic materials used for window glazing, rooflights and lighting diffusers within suspended ceilings, which may not comply with the criteria specified in paragraphs 11 onwards. They are described in the guidance on requirements B2 and B4.

20 For the purposes of the requirements B2 and B4 thermoplastic materials should either be used according to their classification 0-3, under the BS 476: Parts 6 and 7 tests as described in paragraphs 11 onwards, (if they have such a rating), or they may be classified TP(a) rigid, TP(a) flexible, or TP(b) according to the following methods:

TP(a) rigid:

i. rigid solid PVC sheet;

ii. solid (as distinct from double- or multiple-skin) polycarbonate sheet at least 3mm thick;

iii. multi-skinned rigid sheet made from unplasticised PVC or polycarbonate which has a Class 1 rating when tested to BS 476-7:1971, 1987 or 1997; or

iv. any other rigid thermoplastic product, a specimen of which (at the thickness of the product as put on the market), when tested to BS 2782-0:2004 Method 508A *Rate of burning, Laboratory method,* performs so that the test flame extinguishes before the first mark and the duration of flaming or afterglow does not exceed 5 seconds following removal of the burner.

TP(a) flexible:

Flexible products not more than 1mm thick which comply with the Type C requirements of BS 5867-2:1980 *Specification for fabrics for curtains and drapes – Flammability requirements* when tested to BS 5438:1989 *Methods of test for flammability of textile fabrics when subjected to a small igniting flame applied to the face or bottom edge of vertically oriented specimens,* Test 2, with the flame applied to the surface of the specimens for 5, 15, 20 and 30 seconds respectively, but excluding the cleansing procedure; and

TP(b):

i. rigid solid polycarbonate sheet products less than 3mm thick, or multiple-skin polycarbonate sheet products which do not qualify as TP(a) by test; or

ii. other products which, when a specimen of the material between 1.5 and 3mm thick is tested in accordance with BS 2782-0:2004 Method 508A, has a rate of burning which does not exceed 50mm/minute.

Note: If it is not possible to cut or machine a 3mm thick specimen from the product then a 3mm test specimen can be moulded from the same material as that used for the manufacture of the product.

Note: Currently, no new guidance is possible on the assessment or classification of thermoplastic materials under the European system since there is no generally accepted European test procedure and supporting comparative data.

Fire test methods

21 A guide to the various test methods in BS 476 and BS 2782 is given in PD 6520: *Guide to fire test methods for building materials and elements of construction* (available from the British Standards Institution).

A guide to the development and presentation of fire tests and their use in hazard assessment is given in BS 6336:1998 *Guide to development and presentation of fire tests and their use in hazard assessment.*

Table A1 Specific provisions of test for fire resistance of elements of structure etc

Part of building	Minimum provisions when tested to the relevant part of BS 476 [1] (minutes)			Minimum provisions when tested to the relevant European standard (minutes) [9]	Method of exposure
	Loadbearing capacity [2]	Integrity	Insulation		
1. **Structural** frame, beam or column.	See Table A2	Not applicable	Not applicable	R see Table A2	Exposed faces
2. **Loadbearing wall** (which is not also a wall described in any of the following items).	See Table A2	Not applicable	Not applicable	R see Table A2	Each side separately
3. **Floors** [3]					
a. between a shop and flat above;	60 or see Table A2 (whichever is greater)	60 or see Table A2 (whichever is greater)	60 or see Table A2 (whichever is greater)	REI 60 or see Table A2 (whichever is greater)	From underside [4]
b. any other floor – including compartment floors;	See Table A2	See Table A2	See Table A2	REI see Table A2	
c. floor dividing building in different tenancies (excluding flats);	Minimum of 90	Minimum of 90	Minimum of 90	Minimum of REI 90	
d. raised floor and all supports (see note to 7.8).	30				
4. **Roofs**					
a. any part forming an escape route;	30	30	30	REI 30	From underside [4]
b. any roof that performs the function of a floor.	See Table A2	See Table A2	See Table A2	REI see Table A2	
5. **External walls**					
a. any part less than 1000mm from any point on the relevant boundary; [5]	See Table A2 See Table A9	See Table A2 See Table A9	See Table A2 See Table A9	REI see Table A2 See Table A9	Each side separately
b. any part 1000mm or more from the relevant boundary; [5]	See Table A2 See Table A9	See Table A2 See Table A9	15 See Table A9	RE see Table A2 and REI 15 See Table A9	From inside the building
c. any part adjacent to an external escape route (see Section 5, Diagram 25).	30	30	No provision [6][7]	RE 30	From inside the building
6. **Compartment walls** separating a. a flat from any other part of the building (see 8.13); b. occupancies (see 8.18f);	60 or see Table A2 (whichever is less)	60 or see Table A2 (whichever is less)	60 or see Table A2 (whichever is less)	REI 60 or seeTable A2 (whichever is less)	Each side separately
c. buildings in different ownership.	Minimum of 90	Minimum of 90	Minimum of 90	Minimum of REI 90	Each side separately
7a. **Compartment walls** (other than in item 6)	See Table A2	See Table A2	See Table A2	REI see Table A2	Each side separately
8. **Protected shafts** excluding any firefighting shaft					
a. any glazing described in Section 8, Diagram 32;	Not applicable	30	No provision [7]	E 30	Each side separately
b. any other part between the shaft and a protected lobby/corridor described in Diagram 32 above;	30	30	30	REI 30	
c. any part not described in (a) or (b) above.	See Table A2	See Table A2	See Table A2	REI see Table A2	

Table A1 continued

Part of building	Minimum provisions when tested to the relevant part of BS 476 [1] (minutes)			Minimum provisions when tested to the relevant European standard (minutes) [9]	Method of exposure
	Loadbearing capacity [2]	Integrity	Insulation		
9. **Enclosure** (which does not form part of a compartment wall or a protected shaft) to a:					
a. protected stairway;	30	30	30 [8]	REI 30 [8]	Each side separately
b. lift shaft.	30	30	30	REI 30	
10. **Firefighting shafts**	120	120	120	REI 120	From side remote from shaft
a. construction separating firefighting shaft from rest of building;	60	60	60	REI 60	From shaft side
b. construction separating firefighting stair, firefighting lift shaft and firefighting lobby	60	60	60	REI 60	Each side separately
11. **Enclosure** (which is not a compartment wall or described in item 8) to a:					Each side separately
a. protected lobby;	30	30	30 [8]	REI 30 [8]	
b. protected corridor.	30	30	30 [8]	REI 30 [8]	
12. **Sub-division of a corridor**	30	30	30 [8]	REI 30 [8]	Each side separately
13. **Fire-resisting construction:**					Each side separately
a. enclosing places of special fire hazard (see 8.12);	30 (min 60 recommended by insurers)	30 (min 60 recommended by insurers)	30 (min 60 recommended by insurers)	REI 30 30 (min 60 recommended by insurers)	
b. between store rooms and sales area in shops (see 5.58)	30 (min 60 recommended by insurers)	30 (min 60 recommended by insurers)	30 (min 60 recommended by insurers)	REI 30 30 (min 60 recommended by insurers)	
c. fire-resisting subdivision described in Section 2, Diagram 16(b)	30	30	30	REI 30	
d. enclosing bedrooms and ancillary accomodation in care homes (see 3.48 and 3.50)	30	30	30	REI 30	
14. **Enclosure** in a flat to a protected entrance hall, or to a protected landing.	30	30	30 [8]	REI 30 [8]	Each side separately
15. **Cavity barrier**	Not applicable	30	15 (30)	E 30 and EI 15 EI 30	Each side separately
16. **Ceiling** Diagram 35	Not applicable	30	30	EI 30	From underside
17. **Duct** described in paragraph 9.16e	Not applicable	30	No provision	E 30	From outside
18. **Casing** around a drainage system described in Section 10, Diagram 38	Not applicable	30	No provision 30 recommended	E 30 EI 30	From outside
19. **Flue walls** described in Section 10, Diagram 39	Not applicable	Half the period specified in Table A2 for the compartment wall/floor	Half the period specified in Table A2 for the compartment wall/floor	EI half the period specified in Table A2 for the compartment wall/floor	From outside
20. **Fire doors**		See Table B1		See Table B1	

Table A1 **continued**

Notes:

1. Part 21 for loadbearing elements, Part 22 for non-loadbearing elements, Part 23 for fire-protecting suspended ceilings, and Part 24 for ventilation ducts. BS 476-8 results are acceptable for items tested or assessed before 1 January 1988.
2. Applies to loadbearing elements only (see B3.ii and Appendix E).
3. Guidance on increasing the fire resistance of existing timber floors is given in BRE Digest 208 *Increasing the fire resistance of existing timber floors* (BRE 1988).
4. A suspended ceiling should only be relied on to contribute to the fire resistance of the floor if the ceiling meets the appropriate provisions given in Table A3.
5. The guidance in Section 12 allows such walls to contain areas which need not be fire-resisting (unprotected areas).
6. Unless needed as part of a wall in item 5a or 5b.
7. Except for any limitations on glazed elements given in Table A4.
8. See Table A4 for permitted extent of uninsulated glazed elements.
9. The National classifications do not automatically equate with the equivalent classifications in the European column, therefore products cannot typically assume a European class unless they have been tested accordingly.
   'R' is the European classification of the resistance to fire performance in respect of loadbearing capacity; 'E' is the European classification of the resistance to fire performance in respect of integrity; and 'I' is the European classification of the resistance to fire performance in respect of insulation.

Table A2 **Minimum periods of fire resistance**

Purpose group of building	Minimum periods of fire resistance (minutes) in a:					
	Basement storey ($) including floor over		Ground or upper storey			
	Depth (m) of a lowest basement		Height (m) of top floor above ground, in a building or separated part of a building			
	More than 10	Not more than 10	Not more than 5	Not more than 18	Not more than 30	More than 30
1. Residential:						
a. Block of flats – not sprinklered	90	60	30*	60**†	90**	Not permitted
– sprinklered	90	60	30*	60**†	90**	120**
b. Institutional	90	60	30* (60)	60 (90)	90 (120)	120# (180)
c. Other residential	90	60	30* (60)	60 (90)	90 (120)	120# (180)
2. Office:						
– not sprinklered	90	60	30* (60)	60 (90)	90 (120)	Not permitted
– sprinklered (2)	60	60	30*	30* (60)	60 (90)	120#
3. Shop and commercial:						
– not sprinklered	90 (120)	60 (120)	60 (120)	60 (120)	90 (120)	Not permitted
– sprinklered (2)	60 (90)	60 (90)	30* (90)	60 (90)	60 (90)	120#
4. Assembly and recreation:						
– not sprinklered	90	60	60	60 (90)	90 (120)	Not permitted
– sprinklered (2)	60	60	30*	60	60	120#
5. Industrial:						
– not sprinklered	120 (240)§	90 (240)§	60 (240)§	90 (240)§	120 (not recommended)	Not permitted
– sprinklered (2)	90 (120)¶	60 (120)¶	30* (120)¶	60 (120)¶	90 (120)¶	120# (180)
6. Storage and other non-residential:						
a. any building or part not described elsewhere: – not sprinklered	120 (240)	90 (240)	60 (240)	90 (240)	120 (not recommended)	Not permitted
– sprinklered (2)	90 (120)	60 (120)	30* (120)	60 (120)	90 (120)	120#
b. car park for light vehicles:						
i. open sided car park (3)	Not applicable	Not applicable	15*+	15*+ (4)	15*+ (4)	60
ii. any other car park	90	60	30*	60	90	120#

Single storey buildings are subject to the periods under the heading "not more than 5". If they have basements, the basement storeys are subject to the period appropriate to their depth.

$ The floor over a basement (or if there is more than 1 basement, the floor over the topmost basement) should meet the provisions for the ground and upper storeys if that period is higher.

* Increased to a minimum of 60 minutes for compartment walls separating buildings.

** Reduced to 30 minutes for any floor within a flat with more than one storey, but not if the floor contributes to the support of the building.

Reduced to 90 minutes for elements not forming part of the structural frame.

+ Increased to 30 minutes for elements protecting the means of escape.

† Refer to paragraph 7.9 regarding the acceptability of 30 minutes in flat conversions.

§ May be reduced to 120 minutes if hazard grading to *LPC Rules for automatic sprinkler installations* is Ordinary hazard group I or II.

¶ May be reduced to 90 minutes if hazard grading to *LPC Rules for automatic sprinkler installations* is Ordinary hazard group I or II.

Notes:

1. Refer to Table A1 for the specific provisions of test.
2. "Sprinklered" means that the building is fitted throughout with an automatic sprinkler system in accordance with paragraph 0.16.
3. The car park should comply with the relevant provisions in the guidance on requirement B3, Section 11.
4. For the purposes of meeting the Building Regulations, the following types of steel elements are deemed to have satisfied the minimum period of fire resistance of 15 minutes when tested to the European test method;:
   i) Beams supporting concrete floors maximum Hp/A=230m-1 operating under full design load.
   ii) Free standing columns, maximum Hp/A=180m-1 operating under full design load.
   iii) Wind bracing and struts, maximum Hp/A~210m-1 operating under full design load.

   Guidance is also available in BS 5950 Structural use of steelwork in building. Part 8 Code of practice for fire resistant design.

Property protection values for fire resistance take into account such factors as financial exposure, which will vary from building to building. Location of building in relation to fire station may also be a factor as well as variable fire load in factories and warehouse buildings.

Application of the fire resistance standards in Table A2:

a. Where one element of structure supports or carries or gives stability to another, the fire resistance of the supporting element should be no less than the minimum period of fire resistance for the other element (whether that other element is loadbearing or not).

   There are circumstances where it may be reasonable to vary this principle, for example:

   i. where the supporting structure is in the open air and is not likely to be affected by the fire in the building; or

   ii. the supporting structure is in a different compartment, with a fire-separating element (which has the higher standard of fire resistance) between the supporting and the separated structure; or

   iii. where a plant room on the roof needs a higher fire resistance than the elements of structure supporting it.

b. Where an element of structure forms part of more than one building or compartment, that element should be constructed to the standard of the greater of the relevant provisions.

c. Where one side of a basement is (due to the slope of the ground) open at ground level, giving an opportunity for smoke venting and access for fire fighting, it may be appropriate to adopt the standard of fire resistance applicable to above-ground structures for elements of structure in that storey.

d. Although most elements of structure in a single storey building may not need fire resistance (see the guidance on requirement B3, paragraph 7.4(a)), fire resistance will be needed if the element:

   i. is part of (or supports) an external wall and there is provision in the guidance on requirement B4 to limit the extent of openings and other unprotected areas in the wall; or

   ii. is part of (or supports) a compartment wall, including a wall common to two or more buildings; or

   iii. supports a gallery.

For the purposes of this paragraph, the ground storey of a building which has one or more basement storeys and no upper storeys, may be considered as a single storey building. The fire resistance of the basement storeys should be that appropriate to basements.

Table A3 **Limitations on fire-protecting suspended ceilings (see Table A1, Note 4)**

Height of building or separated part (m)	Type of floor	Provision for fire resistance or floor (minutes)	Description of suspended ceiling
Less than 18	Not compartment	60 or less	Type W, X, Y or Z
	Compartment	less than 60	
		60	Type X, Y or Z
18 or more	any	60 or less	Type Y or Z
No limit	any	More than 60	Type Z

Notes:

1. Ceiling type and description (the change from Types A-D to Types W-Z is to avoid confusion with Classes A-D (European)):
   W. Surface of ceiling exposed to the cavity should be Class 0 or Class 1 (National) or Class C-s3, d2 or better (European).
   X. Surface of ceiling exposed to the cavity should be Class 0 (National) or Class B-s3, d2 or better (European).
   Y. Surface of ceiling exposed to the cavity should be Class 0 (National) or Class B-s3, d2 or better (European). Ceiling should not contain easily openable access panels.
   Z. Ceiling should be of a material of limited combustibility (National) or of Class A2-s3, d2 or better (European) and not contain easily openable access panels. Any insulation above the ceiling should be of a material of limited combustibility (National) or Class A2-s3, d2 or better (European).
2. Any access panels provided in fire protecting suspended ceilings of type Y or Z should be secured in position by releasing devices or screw fixings, and they should be shown to have been tested in the ceiling assembly in which they are incorporated.
3. The National classifications do not automatically equate with the equivalent European classifications, therefore, products cannot typically assume a European class unless they have been tested accordingly.

   When a classification includes 's3, d2', this means that there is no limit set for smoke production and/or flaming droplets/particles.

   From a property protection perspective, suspended ceiling type Z should be specified for all commercial building risks regardless of building height.

Table A4 Limitations on the use of uninsulated glazed elements on escape routes (These limitations do not apply to glazed elements which satisfy the relevant insulation criterion, see Table A1)

Position of glazed element	Maximum total glazed area in parts of a building with access to:			
	A single stairway		More than one stairway	
	Walls	Door leaf	Walls	Door leaf
Flats 1. Within the enclosures of a protected entrance hall or protected landing or within fire-resisting separation shown in Section 2 Diagram 4.	Fixed fanlights only	Unlimited above 1100mm from floor	Fixed fanlights only	Unlimited above 1100mm from floor
General 2. Between residential/sleeping accommodation and a common escape route (corridor, lobby or stair).	Nil	Nil	Nil	Nil
3. Between a protected stairway [(1)] and: a. the accommodation; or b. a corridor which is not a protected corridor. Other than in item 2 above.	Nil	25% of door area	Unlimited above 1100mm [(2)]	50% of door area
4. Between: a. a protected stairway [(1)] and a protected lobby or protected corridor; or b. accommodation and a protected lobby. Other than in item 2 above.	Unlimited above 1100mm from floor	Unlimited above 100mm from floor	Unlimited above 100mm from floor	Unlimited above 100mm from floor
5. Between the accommodation and a protected corridor forming a dead end. Other than in item 2 above.	Unlimited above 1100mm from floor	Unlimited above 100mm from floor	Unlimited above 1100mm from floor	Unlimited above 100mm from floor
6. Between accommodation and any other corridor; or subdividing corridors. Other than in item 2 above.	Not applicable	Not applicable	Unlimited above 100mm from floor	Unlimited above 100mm from floor
7. Adjacent an external escape route described in para 3.30.	Unlimited above 1100mm from floor	Unlimited above 1100mm from floor	Unlimited above 1100mm from floor	Unlimited above 1100mm from floor
8. Adjacent an external escape stair (see para 5.25 & Diagram 25) or roof escape (see para 5.35).	Unlimited	Unlimited	Unlimited	Unlimited

Notes:

1. If the protected stairway is also a protected shaft (see paragraph 8.35) or a firefighting stair (see Section 17) there may be further restrictions on the uses of glazed elements.
2. Measured vertically from the landing floor level or the stair pitch line.
3. The 100mm limit is intended to reduce the risk of fire spread from a floor covering.
4. Items 1 and 4 apply also to single storey buildings.
5. Fire-resisting glass should be marked with the manufacturer and product name.
6. Further guidance can be found in A guide to best practice in the specification and use of fire-resistant glazed systems published by the Glass and Glazing Federation.

Table A4a Limitations on the use of uninsulated 'fire-resisting' glazed elements enclosing areas of specific risk (These limitations do not apply to insulated glazed elements)

Position of glazed element	Maximum total glazed area in parts of a building	
	Walls	Door leaf
A. Areas of special risk B. Protected shaft C. Compartment walls	No uninsulated glazing allowed in these applications	25% of door area above 1500mm from floor
D. Individual rooms	Unlimited above 1500mm from floor	Unlimited above 1500mm from floor

Fire resistance capabilities are dependant on a number of factors, including type of fire-resistant glass, type of framing and method of retaining glass in position. Any fire resistance test data should be relevant to the system actually installed as even small changes may reduce the fire resistance achieved.

Table A5 **Notional designations of roof coverings**

Part i: Pitched roofs covered with slates or tiles

Covering material	Supporting structure	Designation
1. Natural slates 2. Fibre reinforced cement slates 3. Clay tiles 4. Concrete tiles	Timber rafters with or without underfelt, sarking, boarding, woodwool slabs, compressed straw slabs, plywood, wood chipboard, or fibre insulating board	AA (National class) or $B_{ROOF}(t4)$ (European class)

Note: Although the Table does not include guidance for roofs covered with bitumen felt, it should be noted that there is a wide range of materials on the market and information on specific products is readily available from manufacturers.

Part ii: Pitched roofs covered with self-supporting sheet

Roof covering material	Construction	Supporting structure	Designation
1. Profiled sheet of galvanised steel, aluminium, fibre reinforced cement, or pre-painted (coil coated) steel or aluminium with a pvc or pvf2 coating	Single skin without underlay, or with underlay or plasterboard, fibre insulating board, or woodwool slab	Structure of timber, steel or concrete	AA (National class) or $B_{ROOF}(t4)$ (European class)
2. Profiled sheet of galvanised steel, aluminium, fibre reinforced cement, or pre-painted (coil coated) steel or aluminium with a pvc or pvf2 coating	Double skin without interlayer, or with interlayer of resin bonded glass fibre, mineral wool slab, polystyrene, or polyurethane	Structure of timber, steel or concrete	AA (National class) or $B_{ROOF}(t4)$ (European class)

Part iii. Flat roofs covered with bitumen felt

A flat roof comprising bitumen felt should (irrespective of the felt specification) be deemed to be of designation AA (National class) or $B_{ROOF}(t4)$ (European class) if the felt is laid on a deck constructed of 6mm plywood, 12.5mm wood chipboard, 16mm (finished) plain edged timber boarding, compressed straw slab, screeded wood wool slab, profiled fibre reinforced cement or steel deck (single or double skin) with or without fibre insulating board overlay, profiled aluminium deck (single or double skin) with or without fibre insulating board overlay, or concrete or clay pot slab (insitu or pre cast), and has a surface finish of:

a. bitumen-bedded stone chippings covering the whole surface to a depth of at least 12.5mm;

b. bitumen-bedded tiles of a non-combustible material;

c. sand and cement screed; or

d. macadam.

Part iv. Pitched or flat roofs covered with fully supported material

Covering material	Supporting structure	Designation
1. Aluminium sheet 2. Copper sheet 3. Zinc sheet 4. Lead sheet 5. Mastic asphalt 6. Vitreous enamelled steel 7. Lead/tin alloy coated steel sheet 8. Zinc/aluminium alloy coated steel sheet 9. Pre-painted (coil coated) steel sheet including liquid-applied pvc coatings	timber joists and: tongued and grooved boarding, or plain edged boarding	AA* (National class) or $B_{ROOF}(t4)$ (European class)
	steel or timber joists with deck of: woodwool slabs, compressed straw slab, wood chipboard, fibre insulating board, or 9.5mm plywood	AA (National class) or $B_{ROOF}(t4)$ (European class)
	concrete or clay pot slab (in-situ or pre-cast) or non-combustible deck of steel, aluminium, or fibre cement (with or without insulation)	AA (National class) or $B_{ROOF}(t4)$ (European class)

Notes:

* Lead sheet supported by timber joists and plain edged boarding should be regarded as having a BA designation and is deemed to be designated class $C_{ROOF}(t4)$ (European class).

The National classifications do not automatically equate with the equivalent classifications in the European column; therefore, products cannot typically assume a European class unless they have been tested accordingly.

Table A6 **Use and definitions of non-combustible materials**

References in AD B guidance to situations where such materials should be used	Definitions of non-combustible materials	
	National class	**European class**
1. refuse chutes meeting the provisions in the guidance to B3, paragraph 8.34c. 2. suspended ceilings and their supports where there is provision in the guidance to B3, paragraph 9.12, for them to be constructed of non-combustible materials. 3. pipes meeting the provisions in the guidance to B3, Table 14. 4. flue walls meeting the provisions in the guidance to B3, Diagram 39. 5. construction forming car parks referred to in the guidance to B3, paragraph 11.3.	a. Any material which when tested to BS 476-11:1982 does not flame nor cause any rise in temperature on either the centre (specimen) or furnace thermocouples b. Totally inorganic materials such as concrete, fired clay, ceramics, metals, plaster and masonry containing not more than 1% by weight or volume of organic material. (Use in buildings of combustible metals such as magnesium/aluminium alloys should be assessed in each individual case). c. Concrete bricks or blocks meeting BS EN 771-1:2003 d. Products classified as non-combustible under BS 476-4:1970	a. Any material classified as class A1 in accordance with BS EN 13501-1:2002 *Fire classification of construction products and building elements*, Part 1 *– Classification using data from reaction to fire tests*. b. Products made from one or more of the materials considered as Class A1 without the need for testing as defined in Commission Decision 2003/424/EC of 6th June 2003 amending Decision 96/603/EC establishing the list of products belonging to Classes A1 "No contribution to fire" provided for in the Decision 94/611/EC implementing Article 20 of the Council Directive 89/106/EEC on construction products. None of the materials shall contain more than 1% by weight or volume (whichever is the more onerous) of homogeneously distributed organic material.

Note:

The National classifications do not automatically equate with the equivalent classifications in the European column, therefore products cannot typically assume a European class unless they have been tested accordingly.

Table A7 **Use and definitions of materials of limited combustibility**

References in AD B guidance to situations where such materials should be used	Definitions of materials of limited combustibility	
	National class	**European class**
1. stairs where there is provision in the guidance to B1 for them to be constructed of materials of limited combustibility (see 5.19). 2. materials above a suspended ceiling meeting the provisions in the guidance to B3, paragraph 9.12. 3. reinforcement/support for fire-stopping referred to in the guidance to B3, see 10.18. 4. roof coverings meeting provisions: a. in the guidance to B3, paragraph 8.29; or b. in the guidance to B4, Table 16 or c. in the guidance to B4, Diagram 47. 5. roof deck meeting the provisions of the guidance to B3, Diagram 30a. 6. class 0 materials meeting the provisions in Appendix A, paragraph 13(a). 7. ceiling tiles or panels of any fire protecting suspended ceiling (Type Z) in Table A3.	a. Any non-combustible material listed in Table A6. b. Any material of density 300/kg/m' or more, which when tested to BS 476-11:1982, does not flame and the rise in temperature on the furnace thermocouple is not more than 20°C. c. Any material with a non-combustible core at least 8mm thick having combustible facings (on one or both sides) not more than 0.5mm thick. (Where a flame spread rating is specified, these materials must also meet the appropriate test requirements).	a. Any material listed in Table A6. b. Any material/product classified as Class A2-s3, d2 or better in accordance with BS EN 13501-1:2002 *Fire classification of construction products and building elements*, Part 1 – *Classification using data from reaction to fire tests*.
8. insulation material in external wall construction referred to in paragraph 12.7. 9. insulation above any fire-protecting suspended ceiling (Type Z) in Table A3.	Any of the materials (a), (b) or (c) above, or: d. Any material of density less than 300kg/m^3, which when tested to BS 476-11:1982, does not flame for more than 10 seconds and the rise in temperature on the centre (specimen) thermocouple is not more than 35°C and on the furnace thermocouple is not more than 25°C.	Any of the materials/products (a) or (b) above.

Note:

1. The National classifications do not automatically equate with the equivalent classifications in the European column; therefore, products cannot typically assume a European class unless they have been tested accordingly.
2. When a classification includes "s3, d2", this means that there is no limit set for smoke production and/or flaming droplets/particles.

Table A8 **Typical performance ratings of some generic materials and products**

Rating	Material or product
Class 0 (National)	1. Any non-combustible material or material of limited combustibility. (Composite products listed in Table A7 must meet test requirements given in Appendix A, paragraph 13(b)). 2. Brickwork, blockwork, concrete and ceramic tiles. 3. Plasterboard (painted or not with a PVC facing not more than 0.5mm thick) with or without an air gap or fibrous or cellular insulating material behind. 4. Woodwool cement slabs. 5. Mineral fibre tiles or sheets with cement or resin binding.
Class 3 (National)	6. Timber or plywood with a density greater than 400kg/m^3, painted or unpainted. 7. Wood particle board or hardboard, either untreated or painted. 8. Standard glass reinforced polyesters.
Class A1 (European)	9. Any material that achieves this class or is defined as 'classified without further test' in a published Commission Decision.
Class A2-s3, d2 (European)	10. Any material that achieves this class or is defined as 'classified without further test' in a published Commission Decision.
Class B-s3, d2 (European)	11. Any material that achieves this class or is defined as 'classified without further test' in a published Commission Decision.
Class C-s3, d2 (European)	12. Any material that achieves this class or is defined as 'classified without further test' in a published Commission Decision.
Class D-s3, d2 (European)	13. Any material that achieves this class or is defined as 'classified without further test' in a published Commission Decision.

Notes (National):

1. Materials and products listed under Class 0 also meet Class 1.
2. Timber products listed under Class 3 can be brought up to Class 1 with appropriate proprietary treatments.
3. The following materials and products may achieve the ratings listed below. However, as the properties of different products with the same generic description vary, the ratings of these materials/products should be substantiated by test evidence.

   Class 0 – aluminium faced fibre insulating board, flame retardant decorative laminates on a calcium silicate board, thick polycarbonate sheet, phenolic sheet and UPVC.

   Class 1 – phenolic or melamine laminates on a calcium silicate substrate and flame-retardant decorative laminates on a combustible substrate.

Notes (European):

For the purposes of the Building Regulations:

1. Materials and products listed under Class A1 also meet Classes A2-s3, d2, B-s3, d2, C-s3, d2 and D-s3, d2.
2. Materials and products listed under Class A2-s3, d2 also meet Classes B-s3, d2, C-s3, d2 and D-s3, d2.
3. Materials and products listed under Class B-s3, d2 also meet Classes C-s3, d2 and D-s3, d2.
4. Materials and products listed under Class C-s3, d2 also meet Class D-s3, d2.
5. The performance of timber products listed under Class D-s3, d2 can be improved with appropriate proprietary treatments.
6. Materials covered by the CWFT process (classification without further testing) can be found by accessing the European Commission's website via the link on the CLG website www.communities.gov.uk
7. The national classifications do not automatically equate with the equivalent classifications in the European column, therefore products cannot typically assume a European class unless they have been tested accordingly.
8. When a classification includes 's3, d2', this means that there is no limit set for smoke production and/or flaming droplets/particles.

Table A9 Property protection from arson attack – specific provisions of test for fire resistance (Method of exposure – fire from outside)

Part of building	Minimum provisions when tested to the relevant part of BS 476:20-22 (minutes)		Minimum provisions when tested to the relevant European standard (minutes)
	Integrity	Insulation	
1. External walls			
a. Unsprinklered buildings			
(i) Within protected zone			
Low and medium risk	60	30	E 60 EI 30
High risk	60	60	EI 60
(ii) Not in the protected zone			
Low and medium risk	30	15	E 30 EI 15
High risk	30	30	EI 30
b. Sprinklered buildings			
(i) Within protected zone			
All risks	60	30	E 60 EI 30
(ii) Not in the protected zone			
All risks	30*	15*	E 30 EI 15
2. Roofs			
a. Unsprinklered buildings			
(i) Within protected zone			
Low and medium risk	60	15	E 60 EI 15
High risk	60	30	EI 30
(ii) Not in the protected zone			
All risks	30	15	E 30 EI 15
b. Sprinklered buildings			
(i) Within protected zone			
All risks	60	15	E 60 EI 15
(ii) Not in the protected zone			
All risks	♣	♣	

Constructions used for the external fabric of a warehouse will be able to be assessed against Table A9 on the basis of prior testing against the appropriate part of BS 476, relevant European Standards and on the basis of historical evidence.

Built-up cladding and sandwich panel systems may also be assessed against Table A9 on the basis of prior testing and approval against LPS 1181-1:Issue 1.1:2005 Elements that achieve grade EXT-B may be assumed to provide no integrity or insulation, but crucially make no significant contribution to fire growth during the developing stages of a fire. Where periods of integrity and insulation are specified in Table A9, appropriate elements of grade EXT-A should be used. (EXT A15 or EXT A30 or EXT A60.)

* EXT B may be acceptable if there is adequate control against arson and pallets etc are at least 10m away from the external wall.

♣ Fire resistance not required but roof cladding must be EXT B.

Table A9 is specifically concerned with fire exposure from the outside. Many materials, products and systems used in the construction of the external fabric of buildings will only have been tested against fire exposure from the inside, (for example, LPS 1181-1 tests for reaction to an internal fire). Materials, products and systems which have only been tested against fire exposure on their internal face will be considered to comply with Table 3 only if they are symmetrical (internal and external faces and substrates the same). If the materials, products or systems are asymmetrical, it will be necessary to show separately that they comply with Table A9 against a fire on the external face.

Appendix B: Fire doors

1. All fire doors should have the appropriate performance given in Table B1 either:

a. by their performance under test to BS 476 *Fire tests on building materials and structures,* Part 22 *Methods for determination of the fire resistance of non-loadbearing elements of construction,* in terms of integrity for a period of minutes, e.g. FD30. A suffix (S) is added for doors where restricted smoke leakage at ambient temperatures is needed; or

b. as determined with reference to Commission Decision 2000/367/EC of 3rd May 2000 implementing Council Directive 89/106/EEC as regards the classification of the resistance to fire performance of construction products, construction works and parts thereof. All fire doors should be classified in accordance with BS EN 13501-2:xxxx, *Fire classification of construction products and building elements. Classification using data from fire resistance tests (excluding products for use in ventilation systems).* They are tested to the relevant European method from the following:

- BS EN 1634-1:2000, *Fire resistance tests for door and shutter assemblies. Fire doors and shutters;*
- BS EN 1634-2:xxxx *Fire resistance tests for door and shutter assemblies. Fire door hardware;*
- BS EN 1634-3:2001 *Fire resistance tests for door and shutter assemblies. Smoke control doors.*

The performance requirement is in terms of integrity (E) for a period of minutes. An additional classification of Sa is used for all doors where restricted smoke leakage at ambient temperatures is needed.

The requirement (in either case) is for test exposure from each side of the door separately, except in the case of lift doors which are tested from the landing side only.

Any test evidence used to substantiate the fire resistance rating of a door or shutter should be carefully checked to ensure that it adequately demonstrates compliance and is applicable to the **complete installed assembly.** Small differences in detail (such as glazing apertures, intumescent strips, door frames and ironmongery etc) may significantly affect the rating.

Note 1: The designation of xxxx is used for standards that are not yet published. The latest version of any standard may be used provided that it continues to address the relevant requirements of the Regulations.

Note 2: Until such time that the relevant harmonised product standards are published, for the purposes of meeting the Building Regulations, products tested in accordance with BS EN 1634-1 (with or without pre-fire test mechanical conditioning) will be deemed to have satisfied the provisions provided that they achieve the minimum fire resistance in terms of integrity, as detailed in Table B1.

2. All fire doors should be fitted with a self-closing device except for fire doors to cupboards and to service ducts which are normally kept locked shut and fire doors within flats (self-closing devices are still necessary on flat entrance doors).

Note: All rolling shutters should be capable of being opened and closed manually for firefighting purposes (see Section 17, paragraph 17.15).

3. Where a self-closing device would be considered a hindrance to the normal approved use of the building, self-closing fire doors may be held open by:

a. a fusible link (but not if the door is fitted in an opening provided as a means of escape unless it complies with paragraph 4 below); or

b. an automatic release mechanism actuated by an automatic fire detection and alarm system; or

c. a door closer delay device.

4. Two fire doors may be fitted in the same opening so that the total fire resistance is the sum of their individual fire resistances, provided that each door is capable of closing the opening. In such a case, if the opening is provided as a means of escape, both doors should be self-closing, but one of them may be fitted with an automatic self-closing device and be held open by a fusible link if the other door is capable of being easily opened by hand and has at least 30 minutes fire resistance.

5. Because fire doors often do not provide any significant insulation, there should be some limitation on the proportion of doorway openings in compartment walls. Therefore no more than 25% of the length of a compartment wall should consist of door openings, unless the doors provide both integrity and insulation to the appropriate level (see Appendix A, Table A2).

Note: Where it is practicable to maintain a clear space on both sides of the doorway, then the above percentage may be greater.

6. Roller shutters across a means of escape should only be released by a heat sensor, such as a fusible link or electric heat detector, in the immediate vicinity of the door. Closure of shutters in such locations should not be initiated by smoke detectors or a fire alarm system, **unless** the shutter is also intended to partially descend to form part of a boundary to a smoke reservoir.

7. Unless shown to be satisfactory when tested as part of a fire door assembly, the essential components of any hinge on which a fire door is hung should be made entirely from materials having a melting point of at least 800°C.

8. Except for doors identified in paragraph 9 below, all fire doors should be marked with the appropriate fire safety sign complying with BS 5499-5:2002 according to whether the door is:

a. to be kept closed when not in use (Fire door keep shut);

b. to be kept locked when not in use (Fire door keep locked shut); or

c. held open by an automatic release mechanism or free swing device (Automatic fire door keep clear).

Fire doors to cupboards and to service ducts should be marked on the outside; all other fire doors on both sides.

9. The following fire doors are not required to comply with paragraph 8 above:

a. doors to and within flats;

b. bedroom doors in 'Other-residential' premises; and

c. lift entrance/landing doors.

10. Tables A1 and A2 set out the minimum periods of fire resistance for the elements of structure to which performance of some doors is linked. Table A4 sets out limitations on the use of uninsulated glazing in fire doors.

11. BS 8214:1990 gives recommendations for the specification, design, construction, installation and maintenance of fire doors constructed with non-metallic door leaves.

Guidance on timber fire-resisting doorsets, in relation to the new European test standard, may be found in *Timber Fire-Resisting Doorsets: maintaining performance under the new European test standard* published by TRADA.

Guidance for metal doors is given in *Code of practice for fire-resisting metal doorsets* published by the DSMA (Door and Shutter Manufacturers' Association) in 1999.

12. Hardware used on fire doors can significantly affect performance in fire. Notwithstanding the guidance in this Approved Document, guidance is available in *Hardware for fire and escape doors* published by the Builders Hardware Industry Federation and Guild of Architectural Ironmongers.

Table B1 **Provisions for fire doors**

Position of door	Minimum fire resistance of door in terms of integrity (minutes) when tested to BS 476-22 [(1)]	Minimum fire resistance of door in terms of integrity (minutes) when tested to the relevant European standard [(3)]
1. **In a compartment wall separating buildings** Also in a compartment wall separating areas of special risk/ high risk	As for the wall in which the door is fitted, but a minimum of 60 (90)	As for the wall in which the door is fitted, but a minimum of 60 (90)
2. **In a compartment wall:**		
a. If it separates a flat from a space in common use;	FD 30S [(2)]	E 30 Sa [(2)]
b. Enclosing a protected shaft forming a stairway situated wholly or partly above the adjoining ground in a building used for Flats, Other Residential, Assembly and Recreation, or Office purposes;	FD 30S [(2)]	E 30 Sa [(2)]
c. enclosing a protected shaft forming a stairway not described in (b) above;	Half the period of fire resistance of the wall in which it is fitted, but 30 minimum and with suffix S [(2)]	Half the period of fire resistance of the wall in which it is fitted, but 30 minimum and with suffix Sa [(2)]
d. enclosing a protected shaft forming a lift or service shaft;	Half the period of fire resistance of the wall in which it is fitted, but 30 minimum	Half the period of fire resistance of the wall in which it is fitted, but 30 minimum
e. not described in (a), (b), (c) or (d) above.	As for the wall it is fitted in, but add S (2) if the door is used for progressive horizontal evacuation under the guidance to B1	As for the wall it is fitted in, but add Sa [(2)] if the door is used for progressive horizontal evacuation under the guidance to B1
3. **In a compartment floor**	As for the floor in which it is fitted	As for the floor in which it is fitted
4. **Forming part of the enclosures of:**		
a. a protected stairway (except as described in item 9) ; or	FD 30S [(2)]	E 30 Sa [(2)]
b. a lift shaft (see paragraph 5.42b); which does not form a protected shaft in 2(b), (c) or (d) above.	FD 30	E 30
5. **Forming part of the enclosure of:**		
a. a protected lobby approach (or protected corridor) to a stairway;	FD 30S [(2)]	E 30 Sa [(2)]
b. any other protected corridor; or	FD 20S [(2)]	E 20 Sa [(2)]
c. a protected lobby approach to a lift shaft (see paragraph 5.42)	FD 30S [(2)]	E 30 Sa [(2)]
6. **Affording access to an external escape route**	FD 30	E 30
7. **Sub-dividing:**		
a. corridors connecting alternative exits;	FD 20S [(2)]	E 20 Sa [(2)]
b. dead-end portions of corridors from the remainder of the corridor	FD 20S [(2)]	E 20 Sa [(2)]
8. **Any door within a cavity barrier**	FD 30	E 30
9. **Any door forming part of the enclosure to a protected entrance hall or protected landing in a flat;**	FD 20	E 20
10. **Any door forming part of the enclosure**		
a. to a place of special fire risk	FD 30 (FD 60)	E 30 (E 60)
b. to ancillary accomodation in care homes (see paragraph 3.50).	FD 30	E 30

Note:

1. To BS 476-22 (or BS 476-8 subject to paragraph 5 in Appendix A).
2. Unless pressurization techniques complying with BS EN 12101-6:2005 Smoke and heat control systems – Part 6: Specification for pressure differential systems – Kits are used, these doors should also either:
   (a) have a leakage rate not exceeding 3m^3/m/hour (head and jambs only) when tested at 25 Pa under BS 476 *Fire tests on building materials and structures*, Section 31.1 *Methods for measuring smoke penetration through doorsets and shutter assemblies, Method of measurement under ambient temperature conditions*; or
   (b) meet the additional classification requirement of Sa when tested to BS EN 1634-3:2001 *Fire resistance tests for door and shutter assemblies*, Part 3 – *Smoke control doors*.
3. The National classifications do not automatically equate with the equivalent classifications in the European column, therefore products cannot typically assume a European class unless they have been tested accordingly.

Appendix C: Methods of measurement

1. Some form of measurement is an integral part of many of the provisions in this document. Paragraphs 2 to 5 and Diagrams C1 to C7 show how the various forms of measurement should be made.

Occupant capacity

2. The **occupant capacity** of a room, storey, building or part of a building is:

a. the maximum number of persons it is designed to hold; or

b. the number calculated by dividing the area of room or storey(s) (m^2) by a floor space factor (m^2 per person) such as those given in Table C1 for guidance.

Note: 'area' excludes stair enclosures, lifts, sanitary accommodation and any other fixed part of the building structure (but counters and display units, etc. should not be excluded).

Table C1 **Floor space factors** [1]

	Type of accommodation [2][3]	Floor space factor m^2/person
1.	Standing spectator areas, bar areas (within 2m of serving point) similar refreshment areas	0.3
2.	Amusement arcade, assembly hall (including a general purpose place of assembly), bingo hall, club, crush hall, dance floor or hall, venue for pop concert and similar events and bar areas without fixed seating	0.5
3.	Concourse, queuing area or shopping mall [4][5]	0.7
4.	Committee room, common room, conference room, dining room, licensed betting office (public area), lounge or bar (other than in 1 above), meeting room, reading room, restaurant, staff room or waiting room [6]	1.0
5.	Exhibition hall or studio (film, radio, television, recording)	1.5
6.	Skating rink	2.0
7.	Shop sales area [7]	2.0
8.	Art gallery, dormitory, factory production area, museum or workshop	5.0
9.	Office	6.0
10.	Shop sales area [8]	7.0
11.	Kitchen or library	7.0
12.	Bedroom or study-bedroom	8.0
13.	Bed-sitting room, billiards or snooker room or hall	10.0
14.	Storage and warehousing	30.0
15.	Car park	Two persons per parking space

Notes:

1. As an alternative to using the values in the table, the floor space factor may be determined by reference to actual data taken from similar premises. Where appropriate, the data should reflect the average occupant density at a peak trading time of year.
2. Where accommodation is not directly covered by the descriptions given, a reasonable value based on a similar use may be selected.
3. Where any part of the building is to be used for more than one type of accommodation, the most onerous factor(s) should be applied. Where the building contains different types of accommodation, the occupancy of each different area should be calculated using the relevant space factor.
4. Refer to section 4 of BS 5588-10:1991 Code of practice for shopping complexes for detailed guidance on the calculation of occupancy in common public areas in shopping complexes.
5. For detailed guidance on appropriate floor space factors for concourses in sports grounds refer to *"Concourses"* published by the Football Licensing Authority (ISBN: 0 95462 932 9).
6. Alternatively the occupant capacity may be taken as the number of fixed seats provided, if the occupants will normally be seated.
7. Shops excluding those under item 10, but including - supermarkets and department stores (main sales areas), shops for personal services such as hairdressing and shops for the delivery or collection of goods for cleaning, repair or other treatment or for members of the public themselves carrying out such cleaning, repair or other treatment.
8. Shops (excluding those in covered shopping complexes but including department stores) trading predominantly in furniture, floor coverings, cycles, prams, large domestic appliances or other bulky goods, or trading on a wholesale self-selection basis (cash and carry).

Travel distance

3. Travel distance is measured by way of the shortest route which if:

a. there is fixed seating or other fixed obstructions, is along the centre line of the seatways and gangways;

b. it includes a stair, is along the pitch line on the centre line of travel.

Width

4. The width of:

a. a **door (or doorway)** is the clear width when the door is open (see Diagram C1);

b. an **escape route** is the width at 1500mm above floor level when defined by walls or, elsewhere, the minimum width of passage available between any fixed obstructions;

c. a **stair** is the clear width between the walls or balustrades.

Note 1: In the case of escape routes and stairs, handrails and strings which do not intrude more than 100mm into these widths may be ignored (see Diagram C1).

Note 2: The rails used for guiding a stair-lift may be ignored when considering the width of a stair. However, it is important that the chair or carriage is able to be parked in a position that does not cause an obstruction to either the stair or landing.

Diagram C1 **Measurement of door width**

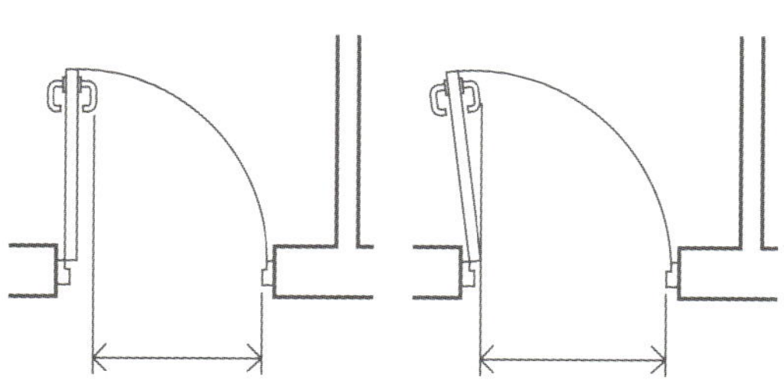

Diagram C2 **Cubic capacity**

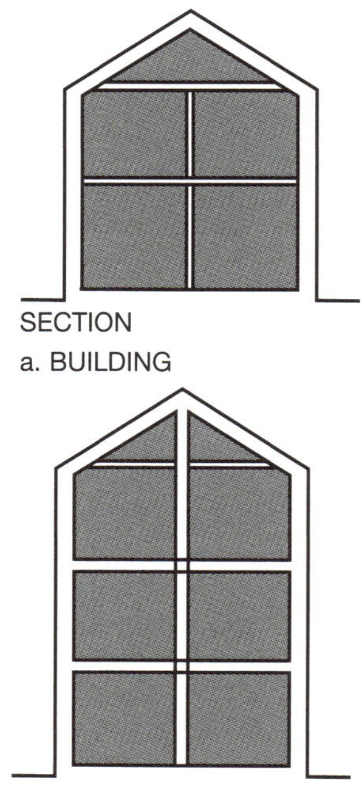

In every case measure the volume contained by:

a under surface of roof as appropriate;

b upper surface of lowest floor;

c the inner surface of enclosing walls (when there is not an enclosing outside wall, measure to the outermost edge of the floor) and ignore internal walls and partitions.

Diagram C3 **Area**

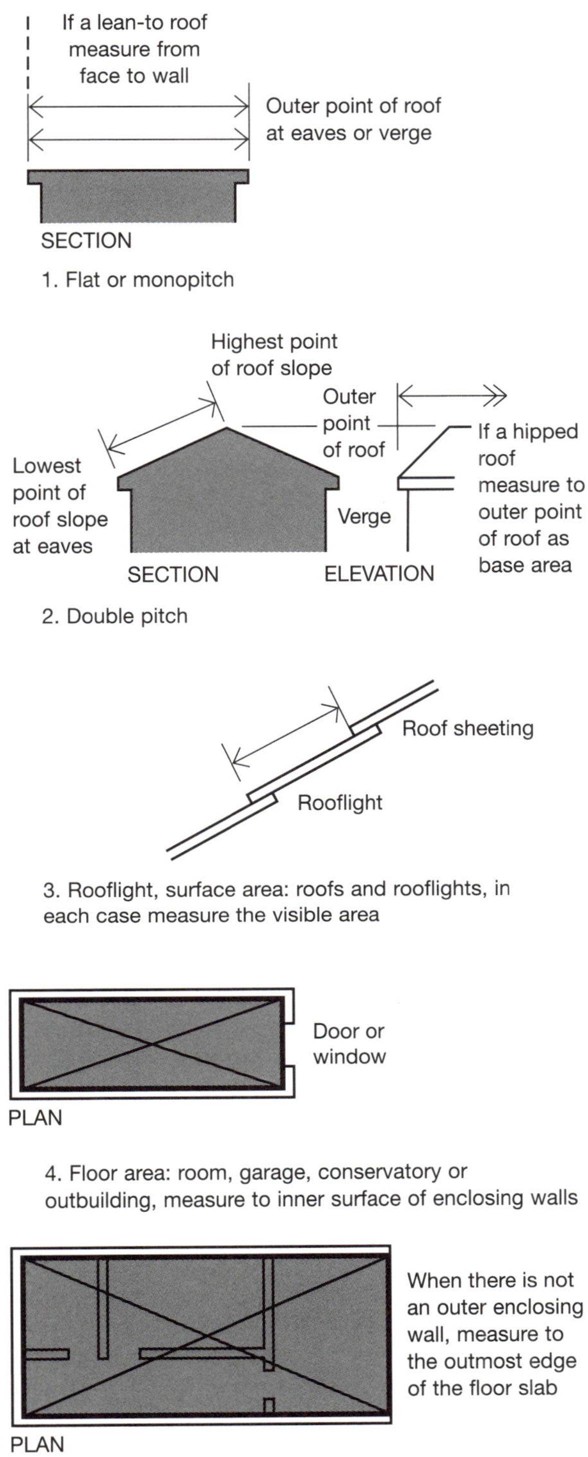

1. Flat or monopitch

2. Double pitch

3. Rooflight, surface area: roofs and rooflights, in each case measure the visible area

4. Floor area: room, garage, conservatory or outbuilding, measure to inner surface of enclosing walls

5. Floor area: storey, part or compartment, measure to inner surface of enclosing walls and include internal walls and partitions

Diagram C4 **Height of building**

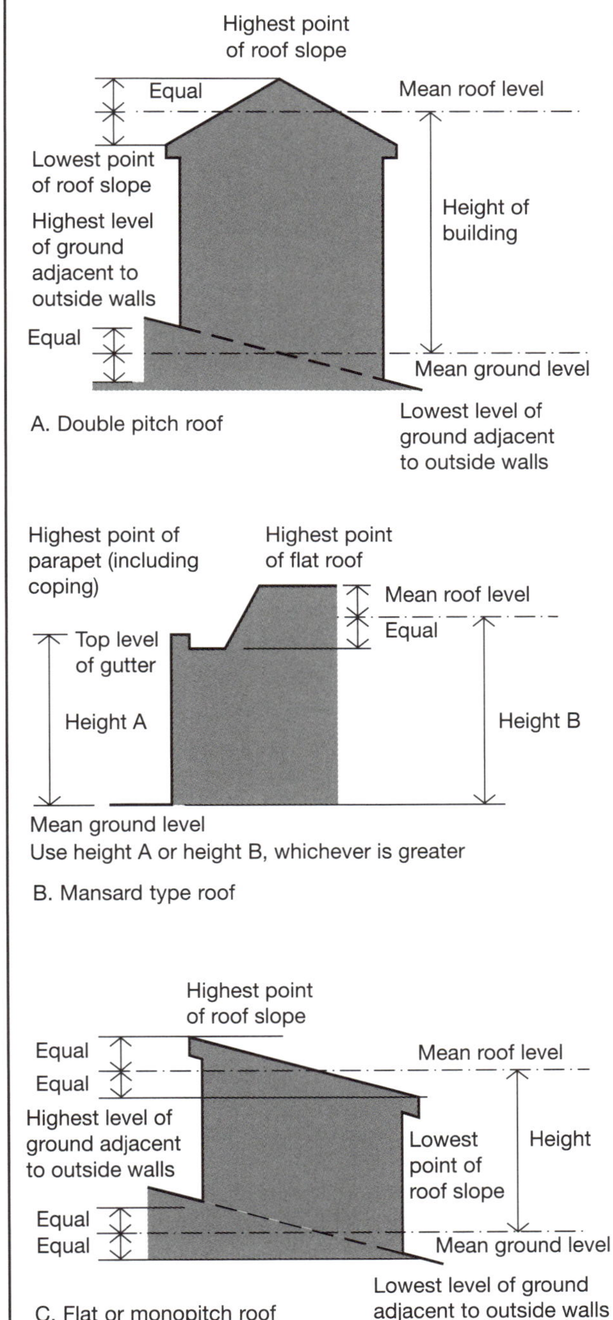

A. Double pitch roof

Use height A or height B, whichever is greater

B. Mansard type roof

C. Flat or monopitch roof

Diagram C5 **Number of storeys**

To count the number of storeys in a building, or in a separated part of a building, count only at the position which gives the greatest number and exclude any basement storeys.

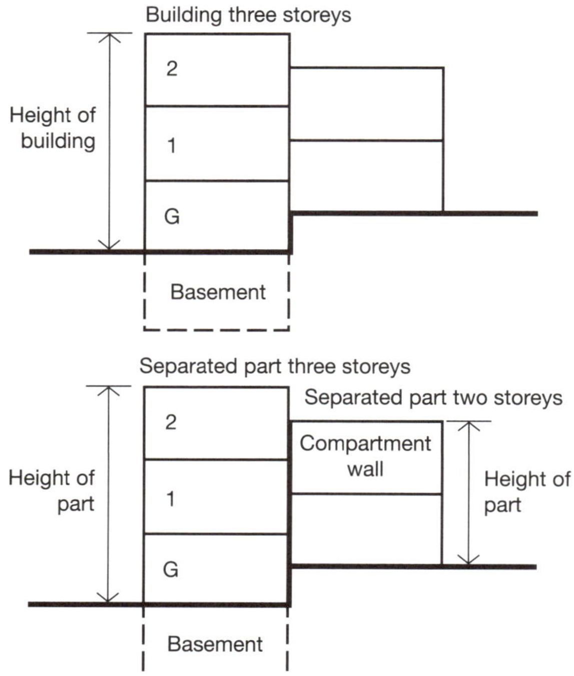

Notes:
In assembly buildings, a gallery is included as a storey, but not if it is a loading gallery, fly gallery, stage grid, lighting bridge, or any gallery provided for similar purposes, or for maintenance and repair.

In other purpose group buildings, galleries are not counted as a storey.

Diagram C6 **Height of top storey in building**

Height of top storey excludes roof-top plant areas and any top storeys consisting exclusively of plant rooms

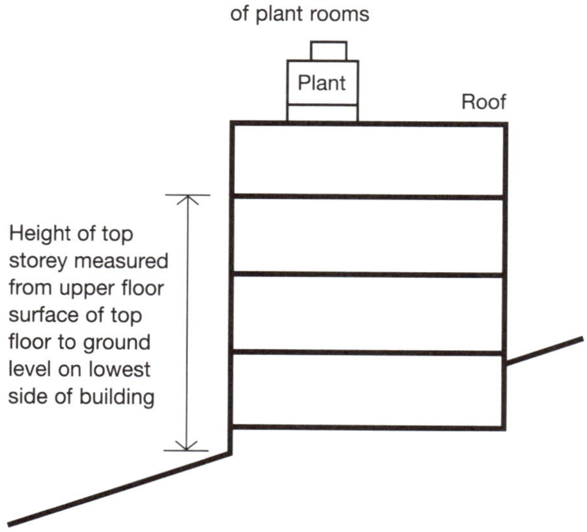

Free Area of Smoke Ventilators

5. The free area of a smoke ventilator, specified in this Approved Document, may be measured by either:

a. the declared aerodynamic free area in accordance with BS EN 12101-2:2003 *Smoke and heat control systems. Specification for natural smoke and heat exhaust ventilators; or,*

b. The total unobstructed cross sectional area, measured in the plane where the area is at a minimum and at right angles to the direction of air flow (see diagram C7).

Diagram C7 **Free area of smoke ventilators**

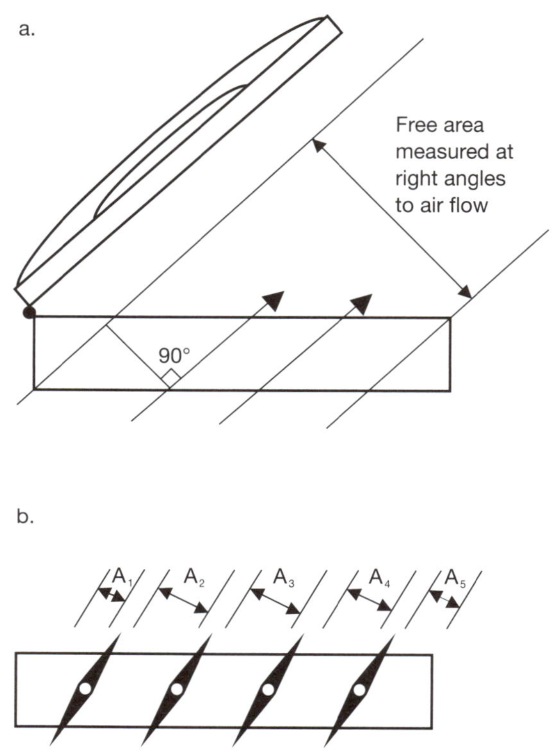

Free area for louvered vent = $A_1+A_2+A_3+A_4+A_5$

Appendix D: Purpose groups

1. Many of the provisions in this document are related to the use of the building. The use classifications are termed purpose groups and represent different levels of hazard. They can apply to a whole building, or (where a building is compartmented) to a compartment in the building and the relevant purpose group should be taken from the main use of the building or compartment.

2. Table D1 sets out the purpose group classification.

Note: This is only of relevance to this Approved Document.

Ancillary and main uses

3. In some situations there may be more than one use involved in a building or compartment and in certain circumstances it is appropriate to treat the different use as belonging to a purpose group in its own right. These situations are:

a. where the ancillary use is a flat; or

b. where the building or compartment is more than 280m^2 in area and the ancillary use is of an area that is more than a fifth of the total floor area of the building or compartment; or

c. storage in a building or compartment of Purpose Group 4 (shop or commercial), where the storage amounts to more than 1/3rd of the total floor area of the building or compartment and the building or compartment is more than 280m^2 in area.

4. Some buildings may have two or more main uses that are not ancillary to one another. For example offices over shops from which they are independent. In such cases, each of the uses should be considered as belonging to a purpose group in its own right.

5. In other cases and particularly in some large buildings, there may be a complex mix of uses. In such cases it is necessary to consider the possible risk that one part of a complex may have on another and special measures to reduce the risk may be necessary.

Table D1 **Classification of Purpose Groups**

Title	Group	Purpose for which the building or compartment of a building is intended to be used
Residential (dwellings)	1(a)*	Flat.
	1(b)†	Dwellinghouse which contains a habitable storey with a floor level which is more than 4.5m above ground level.
	1(c)†+	Dwellinghouse which does not contain a habitable storey with a floor level which is more than 4.5m above ground level.
Residential (Institutional)	2(a)	Hospital, home, school or other similar establishment used as living accommodation for, or for the treatment, care or maintenance of persons suffering from disabilities due to illness or old age or other physical or mental incapacity, or under the age of 5 years, or place of lawful detention, where such persons sleep on the premises.
(Other)	2(b)	Hotel, boarding house, residential college, hall of residence, hostel and any other residential purpose not described above.
Office	3	Offices or premises used for the purpose of administration, clerical work (including writing, book keeping, sorting papers, filing, typing, duplicating, machine calculating, drawing and the editorial preparation of matter for publication, police and fire and rescue service work), handling money (including banking and building society work), and communications (including postal, telegraph and radio communications) or radio, television, film, audio or video recording, or performance (not open to the public) and their control.
Shop and commercial	4	Shops or premises used for a retail trade or business (including the sale to members of the public of food or drink for immediate consumption and retail by auction, self-selection and over-the-counter wholesale trading, the business of lending books or periodicals for gain and the business of a barber or hairdresser and the rental of storage space to the public) and premises to which the public is invited to deliver or collect goods in connection with their hire repair or other treatment, or (except in the case of repair of motor vehicles) where they themselves may carry out such repairs or other treatments.
Assembly and recreation	5	Place of assembly, entertainment or recreation; including bingo halls, broadcasting, recording and film studios open to the public, casinos, dance halls; entertainment, conference, exhibition and leisure centres; funfairs and amusement arcades; museums and art galleries; non-residential clubs, theatres, cinemas and concert halls; educational establishments, dancing schools, gymnasia, swimming pool buildings, riding schools, skating rinks, sports pavilions, sports stadia; law courts; churches and other buildings of worship, crematoria; libraries open to the public, non-residential day centres, clinics, health centres and surgeries; passenger stations and termini for air, rail, road or sea travel; public toilets; zoos and menageries.
Industrial	6	Factories and other premises used for manufacturing, altering, repairing, cleaning, washing, breaking-up, adapting or processing any article; generating power or slaughtering livestock.
Storage and other non-residential+	7(a)	Place for the storage or deposit of goods or materials (other than described under 7(b)) and any building not within any of the Purpose Groups 1 to 6.
	7(b)	Car parks designed to admit and accommodate only cars, motorcycles and passenger or light goods vehicles weighing no more than 2500kg gross.

Notes:

This table only applies to Part B.

* Includes live/work units that meet the provisions of paragraph 2.52.

† includes any surgeries, consulting rooms, offices or other accommodation, not exceeding $50m^2$ in total, forming part of a dwellinghouse and used by an occupant of the dwellinghouse in a professional or business capacity.

\+ A detached garage not more than $40m^2$ in area is included in purpose group 1(c); as is a detached open carport of not more than $40m^2$, or a detached building which consists of a garage and open carport where neither the garage nor the open carport exceeds $40m^2$ in area.

Appendix E: Definitions

Note: Except for the items marked * (which are from the Building Regulations), these definitions apply only to Part B.

Access room A room through which passes the only escape route from an inner room.

Accommodation stair A stair, additional to that or those required for escape purposes, provided for the convenience of occupants.

Alternative escape routes Escape routes sufficiently separated by either direction and space, or by fire-resisting construction, to ensure that one is still available should the other be affected by fire.

Alternative exit One of two or more exits, each of which is separate from the other.

Appliance ventilation duct A duct provided to convey combustion air to a gas appliance.

Atrium (plural atria) A space within a building, not necessarily vertically aligned, passing through one or more structural floors.

Note: Enclosed lift wells, enclosed escalator wells, building services' ducts and stairways are not classified as atria.

Automatic release mechanism A device which will allow a door held open by it to close automatically in the event of each or any one of the following:

a. detection of smoke by automatic apparatus suitable in nature, quality and location;

b. operation of a hand-operated switch fitted in a suitable position;

c. failure of electricity supply to the device, apparatus or switch;

d. operation of the fire alarm system if any.

Basement storey A storey with a floor which at some point is more than 1200mm below the highest level of ground adjacent to the outside walls. (However, see Appendix A, Table A2, for situations where the storey is considered to be a basement only because of a sloping site.)

Boundary The boundary of the land belonging to the building, or, where the land abuts a road, railway, canal or river, the centreline of that road, railway, canal or river. (See Diagram 41.)

***Building** Any permanent or temporary building but not any other kind of structure or erection. A reference to a building includes a reference to part of a building.

Building Control Body A term used to include both Local Authority Building Control and Approved Inspectors.

Cavity barrier A construction, other than a smoke curtain, provided to close a concealed space against penetration of smoke or flame, or provided to restrict the movement of smoke or flame within such a space.

Ceiling A part of a building which encloses and is exposed overhead in a room, protected shaft or circulation space. (The soffit of a rooflight is included as part of the surface of the ceiling, but not the frame. An upstand below a rooflight would be considered as a wall).

Circulation space A space (including a protected stairway) mainly used as a means of access between a room and an exit from the building or compartment.

Class 0 A product performance classification for wall and ceiling linings. The relevant test criteria are set out in Appendix A, paragraph 13.

Common balcony A walkway, open to the air on one or more sides, forming part of the escape route from more than one flat.

Common stair An escape stair serving more than one flat.

Compartment (fire) A building or part of a building, comprising one or more rooms, spaces or storeys, constructed to prevent the spread of fire to or from another part of the same building, or an adjoining building. (A roof space above the top storey of a compartment is included in that compartment.) (See also "Separated part".)

Compartment wall or floor A fire-resisting wall/floor used in the separation of one fire compartment from another. (Constructional provisions are given in Section 8).

Concealed space or cavity A space enclosed by elements of a building (including a suspended ceiling) or contained within an element, but not a room, cupboard, circulation space, protected shaft or space within a flue, chute, duct, pipe or conduit.

Corridor access A design of a building containing flats in which each flat is approached via a common horizontal internal access or circulation space which may include a common entrance hall.

Dead end Area from which escape is possible in one direction only.

Direct distance The shortest distance from any point within the floor area, measured within the external enclosures of the building, to the nearest storey exit ignoring walls, partitions and fittings, other than the enclosing walls/partitions to protected stairways.

Dwelling A unit of residential accommodation occupied (whether or not as a sole or main residence):

a. by a single person or by people living together as a family; or

b. by not more than 6 residents living together as a single household, including a household where care is provided for residents.

Element of structure

a. a member forming part of the structural frame of a building or any other beam or column;

b. a loadbearing wall or loadbearing part of a wall;

c. a floor;

d. a gallery (but not a loading gallery, fly gallery, stage grid, lighting bridge, or any gallery provided for similar purposes or for maintenance and repair);

e. an external wall; and

f. a compartment wall (including a wall common to two or more buildings). (However, see the guidance to B3, paragraph 7.4, for exclusions from the provisions for elements of structure.)

Emergency lighting Lighting provided for use when the supply to the normal lighting fails.

Escape lighting That part of the emergency lighting which is provided to ensure that the escape route is illuminated at all material times.

Escape route Route forming that part of the means of escape from any point in a building to a final exit.

European Technical Approval A favourable technical assessment of the fitness for use of a construction product for an intended use, issued for the purposes of the Construction Products Directive by a body authorised by a member State to issue European Technical Approvals for those purposes and notified by that member State to the European Commission.

European Technical Approvals Issuing body A body notified under Article 10 of the Construction Products Directive. The details of these institutions are published in the "C" series of the Official Journal of the European Communities.

Evacuation lift A lift that may be used for the evacuation of people in a fire.

Exit passageway A protected passageway connecting a protected stairway to a final exit (exit passageways should be protected to the same standard as the stairway they serve).

External wall (or side of a building) Includes a part of a roof pitched at an angle of more than 70° to the horizontal, if that part of the roof adjoins a space within the building to which persons have access (but not access only for repair or maintenance).

Final exit The termination of an escape route from a building giving direct access to a street, passageway, walkway or open space and sited to ensure the rapid dispersal of persons from the vicinity of a building so that they are no longer in danger from fire and/or smoke.

Note: Windows are not acceptable as final exits.

Fire damper Mechanical or intumescent device within a duct or ventilation opening which is operated automatically and is designed to prevent the passage of fire and which is capable of achieving an integrity E classification and/or an ES classification to BS EN13501-3:2005 when tested to BS EN1366-2:1999. Intumescent fire dampers may be tested to ISO 10294-5.

Fire and smoke damper Fire damper which when tested in accordance with BS EN 1366-2:1999 meets the ES classification requirements defined in EN 13501-3:2005 and achieves the same fire resistance in relation to integrity, as the element of the building construction through which the duct passes. Intumescent fire dampers may be tested to ISO 10294-2.

Fire door A door or shutter, provided for the passage of persons, air or objects, which, together with its frame and furniture as installed in a building, is intended (when closed) to resist the passage of fire and/or gaseous products of combustion and is capable of meeting specified performance criteria to those ends. (It may have one or more leaves and the term includes a cover or other form of protection to an opening in a fire-resisting wall or floor, or in a structure surrounding a protected shaft.)

Firefighting lift A lift designed to have additional protection, with controls that enable it to be used under the direct control of the fire and rescue service in fighting a fire. (See Sections 15-17.)

Firefighting lobby A protected lobby providing access from a firefighting stair to the accommodation area and to any associated firefighting lift.

Firefighting shaft A protected enclosure containing a firefighting stair, firefighting lobbies and, if provided, a firefighting lift, together with its machine room.

Firefighting stair A protected stairway communicating with the accommodation area only through a firefighting lobby.

Fire-resisting (fire resistance) The ability of a component or construction of a building to satisfy for a stated period of time, some or all of the appropriate criteria specified in the relevant standard test.

Fire-separating element A compartment wall, compartment floor, cavity barrier and construction enclosing a protected escape route and/or a place of special fire hazard.

Fire stop A seal provided to close an imperfection of fit or design tolerance between elements or components, to restrict the passage of fire and smoke.

* **Flat** A separate and self contained premises constructed or adapted for use for residential purposes and forming part of a building from some other part of which it is divided horizontally.

Gallery A floor or balcony which does not extend across the full extent of a building's footprint and is open to the floor below.

Habitable room A room used, or intended to be used, for dwelling purposes (including for the purposes of Part B, a kitchen, but not a bathroom).

Height (of a building or storey for the purposes of Part B) Height of a building is measured as shown in Appendix C, Diagram C4 and height of the floor of the top storey above ground is measured as shown in Appendix C, Diagram C6.

High risk area Part of a building containing hazardous or unique processes, high value goods or plant or other aspects creating a significant financial exposure.

Inner room Room from which escape is possible only by passing through another room (the access room).

Live/Work Unit A flat which is intended to serve as a workplace for its occupants and for persons who do not live on the premises.

Material of limited combustibility A material performance specification that includes non-combustible materials and for which the relevant test criteria are set out in Appendix A, paragraph 9.

Means of escape Structural means whereby [in the event of fire] a safe route or routes is or are provided for persons to travel from any point in a building to a place of safety.

Measurement Width of a doorway, area, cubic capacity, height of a building and number of storeys, see Appendix C, Diagrams C1 to C7; occupant capacity, travel distance and, escape route and a stair, see Appendix C.

Non-combustible material The highest level of reaction to fire performance. The relevant test criteria are set out in Appendix A, paragraph 8.

Notional boundary A boundary presumed to exist between buildings on the same site (see Section 13, Diagram 42).

Occupancy type A purpose group identified in Appendix D.

Open spatial planning The internal arrangement of a building in which more than one storey or level is contained in one undivided volume, e.g. split-level floors. For the purposes of this document there is a distinction between open spatial planning and an atrium space.

Perimeter (of a building) The maximum aggregate plan perimeter, found by vertical projection onto a horizontal plane (see Section 16, Diagram 48).

Pipe (for the purposes of Section 10) – includes pipe fittings and accessories and excludes a flue pipe and a pipe used for ventilating purposes (other than a ventilating pipe for an above around drainage system).

Places of special fire hazard Oil-filled transformer and switch gear rooms, boiler rooms, storage space for fuel or other highly flammable substances and rooms housing a fixed internal combustion engine.

Platform floor (access or raised floor) A floor supported by a structural floor, but with an intervening concealed space which is intended to house services.

Protected circuit An electrical circuit protected against fire.

Protected corridor/lobby A corridor or lobby which is adequately protected from fire in adjoining accommodation by fire-resisting construction.

Protected entrance hall/landing A circulation area consisting of a hall or space in a flat, enclosed with fire-resisting construction (other than any part which is an external wall of a building).

Protected shaft A shaft which enables persons, air or objects to pass from one compartment to another and which is enclosed with fire-resisting construction.

Protected stairway A stair discharging through a final exit to a place of safety (including any exit passageway between the foot of the stair and the final exit) that is adequately enclosed with fire-resisting construction.

Protected zone The sections of roof, external walls and supporting frame of a single-storey building adjacent to and within a specified distance on each side of a compartment wall.

Purpose group A classification of a building according to the purpose to which it is intended to be put. See Appendix D, Table D1.

Relevant boundary The boundary which the side of the building faces, (and/or coincides with) and which is parallel, or at an angle of not more than 80°, to the side of the building (see Section 13, Diagram 41). A notional boundary can be a relevant boundary.

Rooflight A dome light, lantern light, skylight, ridge light, glazed barrel vault or other element intended to admit daylight through a roof.

Room (for the purposes of B2) An enclosed space within a building that is not used solely as a circulation space. (The term includes not only conventional rooms, but also cupboards that are not fittings and large spaces such as warehouses and auditoria. The term does not include voids such as ducts, ceiling voids and roof spaces.)

School A place of education for children older than 2 and younger than 19 years. Includes nursery schools, primary schools and secondary schools as defined in the Education Act 1996.

Self-closing device A device which is capable of closing the door from any angle and against any latch fitted to the door.

> **Note:** Rising butt hinges which do not meet the above criteria are acceptable where the door is in a cavity barrier.

Separated part (of a building) A form of compartmentation in which a part of a building is separated from another part of the same building by a compartment wall. The wall runs the full height of the part and is in one vertical plane. (See paragraph 8.22 and Appendix C, Diagram C5.)

Sheltered housing includes:

a. two or more dwellings in the same building;

b. two or more dwellings on adjacent sites

where those dwellings are, in each case, designed and constructed for the purpose of providing residential accommodation for vulnerable or elderly people who receive, or who are to receive, a support service.

Single storey building A building consisting of a ground storey only. (A separated part which consists of a ground storey only, with a roof to which access is only provided for repair or maintenance, may be treated as a single storey building). Basements are not included in counting the number of storeys in a building (see Appendix C).

Site (of a building) is the land occupied by the building, up to the boundaries with land in other ownership.

Smoke alarm A device containing within one housing all the components, except possibly the energy source, necessary for detecting smoke and giving an audible alarm.

Special risk Any part of a building, including a compartment or single room, that contains for example plant or goods of high value, high hazard goods or processes, data processing equipment or any business-critical area.

Storey includes:

a. any gallery in an assembly building (Purpose Group 5); and

b. any gallery in any other type of building if its area is more than half that of the space into which it projects; and

   Note: Where there is more than one gallery and the total aggregate area of all the galleries in any one space is more than half of the area of that space then the building should be regarded as being a multi storey building.

c. a roof, unless it is accessible only for maintenance and repair.

Storey exit A final exit, or a doorway giving direct access into a protected stairway, firefighting lobby, or external escape route.

Note: A door in a compartment wall in an institutional building is considered as a storey exit for the purposes of B1 if the building is planned for progressive horizontal evacuation, see paragraph 3.41.

Suspended ceiling (fire-protecting) A ceiling suspended below a floor, which contributes to the fire resistance of the floor. Appendix A, Table A3, classifies different types of suspended ceiling.

Technical specification A standard or a European Technical Approval Guide. It is the document against which compliance can be shown in the case of a standard and against which an assessment is made to deliver the European Technical Approval.

Thermoplastic material See Appendix A, paragraph 17.

Travel distance (unless otherwise specified, e.g. as in the case of flats) The actual distance to be travelled by a person from any point within the floor area to the nearest storey exit, having regard to the layout of walls, partitions and fittings.

Unprotected area In relation to a side or external wall of a building means:

a. window, door or other opening; and

   Note: Windows that are not openable and are designed and glazed to provide the necessary level of fire resistance and recessed car parking areas shown in Diagram E1, need not be regarded as an unprotected area.

b. any part of the external wall which has less than the relevant fire resistance set out in Section 12; and

c. any part of the external wall which has combustible material more than 1mm thick attached or applied to its external face, whether for cladding or any other purpose. (Combustible material in this context is any material which does not have a Class 0 rating.)

Diagram E1 **Recessed car parking areas**

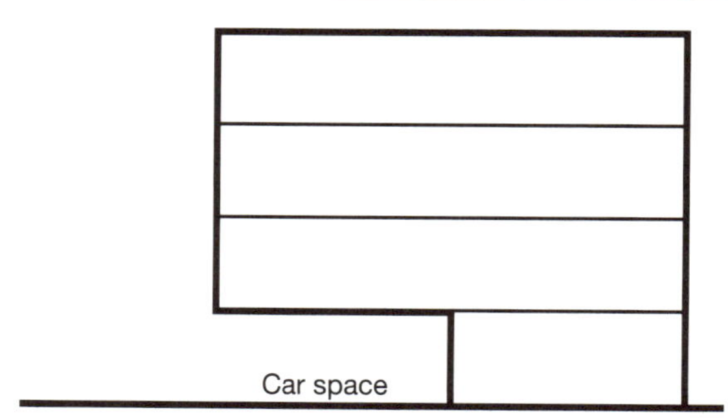

Note: The parking area should be:

a Open fronted

b Separated from the remainder of the building by a compartment wall(s) and floor(s) having not less than the period of fire resistance specified in Table A2 in Appendix A.

Appendix F: Fire behaviour of insulating core panels used for internal structures

1. Insulating core panel systems are used for external cladding as well as for internal structures. However, whilst both types of panel system have unique fire behaviour characteristics, it is those used for internal structures that can present particular problems with regard to fire spread.

The most common use of insulating core panels, when used for internal structures, is to provide an enclosure in which a chilled or sub zero environment can be generated for the production, preservation, storage and distribution of perishable foodstuffs. However this type of construction is also used in many other applications, particularly where the maintenance of a hygienic environment is essential.

These panels typically consist of an inner core sandwiched between and bonded to facings of galvanised steel, often with a PVC facing for hygiene purposes. The panels are then formed into a structure by jointing systems, usually designed to provide an insulating and hygienic performance. The panel structure can be free standing, but is usually attached to the building structure by lightweight fixings or hangers in the case of ceilings.

The most common forms of insulation in present use are:

- polyisocyanurate,
- mineral fibre,
- phenolic,
- polystyrene (cold stores),
- extruded polystyrene,
- composite polymers such as syntactic phenolic.

Fire behaviour of the core materials and fixing systems

2. The degradation of polymeric materials can be expected when exposed to radiated/conducted heat from a fire, with the resulting production of large quantities of smoke.

It is recognised that the potential for problems in fires involving mineral fibre cores is generally less than those for polymeric core materials.

In addition, irrespective of the type of core material, the panel, when exposed to the high temperatures of a developed fire, will tend to delaminate between the facing and core material, due to a combination of expansion of the metal facing and softening of the bond line.

Therefore once it is involved, either directly or indirectly in a fire, the panel will have lost most of its structural integrity. Tability will then be dependant on the method of fixing to the structure. For systems that are not fixed through both facings the stability of the system will then depend on the residual structural strength of the non-exposed facing, the interlocking joint between panels and the fixing system.

Most jointing or fixing systems for these systems have an extremely limited structural integrity performance in developed fire conditions. If the fire starts to heat up the support fixings or structure to which they are attached, then there is a real chance of total collapse of the panel system.

Where panels are used a the lining to a building the insulating nature of these panels, together with their sealed joints, means that fire can spread behind the panels, hidden from the occupants of occupied rooms/spaces. With some thermoplastic cores fire can also spread between the panel facings.

This can prove to be a particular problem to firefighters as, due to the insulating properties of the cores, it may not be possible to track the spread of fire, even using infra red detection equipment. This difficulty, together with that of controlling the fire spread within and behind the panels, is likely to have a detrimental effect on the performance of the fixing systems, potentially leading to their complete and unexpected collapse, together with any associated equipment.

Firefighting

3. When compared with other types of construction techniques, these panel systems therefore provide a unique combination of problems for firefighters, including:

- hidden fire spread within panels with thermoplastic cores;
- production of large quantities of black toxic smoke; and
- rapid fire spread leading to flashover.
- Hidden fire behind lining systems

These three characteristics are common to both polyurethane and polystyrene cored panels, although the rate of fire spread in polyurethane cores is significantly less than that of polystyrene cores, especially when any external heat source is removed.

In addition, irrespective of the type of panel core, all systems are susceptible to:

- delamination of the steel facing;
- collapse of the system; and
- hidden fire spread behind the system.

Design recommendations

4. To identify the appropriate solution, a risk assessment approach should be adopted. This would involve identifying the potential fire risk within the enclosures formed by the panel systems and then adopting one or more of the following at the design stage:

- removing the risk;
- separating the risk from the panels by an appropriate distance;
- providing a fire suppression system for the risk;
- providing a fire suppression system for the enclosure;
- providing fire-resisting panels; and
- specifying appropriate materials/fixing and jointing systems.

In summary the performance of the building structure, including the insulating envelope, the superstructure, the substructure etc, must be considered in relation to their performance in the event of a fire.

Specifying panel core materials

5. Where at all possible the specification of panels with core materials appropriate to the application will help ensure an acceptable level of performance for panel systems, when involved in fire conditions.

The following are examples in the provision of core materials which may be appropriate to the application concerned.

Mineral fibre cores:

- cooking areas,
- hot areas,
- bakeries,
- fire breaks in combustible panels,
- fire stop panels,
- general fire protection.

All cores:

- chill stores,
- cold stores,
- blast freezers,
- food factories,
- clean rooms.

Note: Core materials may be used in other circumstances where a risk assessment has been made and other appropriate fire precautions have been put in place.

Specifying materials/fixing and jointing systems

6. The following are methods by which the stability of panel systems may be improved in the event of a fire, although they may not all be appropriate in every case.

In addition the details of construction of the insulating envelope should, particularly in relation to combustible insulant cores, prevent the core materials from becoming exposed to the fire and contributing to the fire load.

a. Insulating envelopes, support systems and supporting structure should be designed to allow the envelope to remain structurally stable by alternative means such as catenary action following failure of the bond line between insulant core and facing materials. This particularly relates to ceilings and will typically require positive attachment of the lower faces of the insulant panels to supports.

b. The building superstructure, together with any elements providing support to the insulating envelope, should be protected to prevent early collapse of the structure or the envelope.

   Note: Irrespective of the type of panel provided, it will remain necessary to ensure that the supplementary support method supporting the panels remains stable for an appropriate time period under fire conditions. It is not practical to fire protect light gauge steel members such as purlins and sheeting rails which provide stability to building superstructures and these may be compromised at an early stage of a fire. Supplementary fire-protected heavier gauge steelwork members could be provided at wider intervals than purlins to provide restraint in the event of a fire.

c. In designated high risk areas, consideration should be given to incorporating non-combustible insulant cored panels into wall and ceiling construction at intervals, or incorporating strips of non-combustible material into specified wall and ceiling panels, in order to provide a barrier to fire propagation through the insulant.

d. Correct detailing of the insulating envelope should ensure that the combustible insulant is fully encapsulated by non-combustible facing materials which remain in place during a fire.

e. The panels should incorporate pre-finished and sealed areas for penetration of services.

General

7. Generally, panels or panel systems should not be used to support machinery or other permanent loads.

Any cavity created by the arrangement of panels, their supporting structure or other building elements should be provided with suitable cavity barriers.

8. Examples of possible solutions and general guidance on insulating core panels construction can be found in *Design, construction, specification and fire management of insulated envelopes for temperature controlled environments* published by the International Association of Cold Storage Contractors (European Division).

Of particular relevance is Chapter 8 of the document which gives guidance on the design, construction and management of insulated structures. Whilst the document is primarily intended for use in relation to cold storage environments, the guidance, particularly in Chapter 8, is considered to be appropriate for most insulating core panel applications.

Composite panels – insurers' view

Over the last decade or so significant losses have occurred associated with large fires in which composite panels have been a feature. Many have related to risks within the food manufacturing sector.

Important issues need to be considered such as fire safety management, compartmentation, panel system construction, combustibility and fire performance of panel core material and the choice of system for applications with differing inception and fire spread risks.

As with other types of risk, buildings containing sandwich panel systems should be considered on their own merits by underwriters and risk managers, taking account of application, choice of sandwich panel system and fire risk management measures in place. This three-pronged approach is necessary, balancing negative factors in any one area with strengths elsewhere.

Sandwich panels do not start a fire on their own, and where these systems have been implicated in fire spread, the fire has often started in high-risk areas such as cooking areas, subsequently spreading as a result of poor fire risk management, and poor prevention and containment measures. Prevention of ignition and containment of early fire spread are critical. Specific and detailed risk assessment is crucial.

High levels of risk management cannot be assumed by the building designer. Due to the nature of the processes in the business in question, and where the risk of ignition is high, the use of panel systems with high fire performance characteristics should be considered. Systems meeting accreditation schemes such as LPS 1181 demonstrate such characteristics.

Some applications, including stand-alone cold stores and panel systems used as external claddings in areas where arson risk is low, have experienced few fire losses. In low-risk situations such as these there can be greater flexibility in the choice of panel system, taking account of other business needs such as hygienic environments and insulation properties.

Inevitably many situations will fall between clear-cut 'high' or 'low' risk scenarios. Here the degree of financial exposure is likely to drive insurers' decisions. Fire-stop panels and other fire safety management measures have a significant role in such situations. The importance of a demonstrated ability by facilities managers to ensure that such systems are robust cannot be over-emphasised.

The spread of fire is influenced by:

- *fire safety management;*
- *compartmentation;*
- *sandwich panel system construction;*
- *combustibility and fire performance of panel core material;*
- *application of sandwich panel system.*

Not all sandwich panels contain a combustible core. Even sandwich panels containing a combustible core do not themselves create the fire risk; they require a significant size of ignition source to propagate the fire. Not all combustible materials used in the construction of sandwich panels have poor fire performance.

Good standards of fire safety management can substantially reduce the fire risk.

Fire spread via the external envelope is not so common and therefore does not have the same level of risk. Fire statistics confirm this and replacement of such panels is probably not justified. Some reasonable control of fire performance may be needed, particularly for higher risks. It may not be necessary to replace all external envelopes (claddings) installed in existing buildings, provided there are other compensatory risk improvements.

Sandwich panel systems used for external claddings typically have a polyurethane core and are supported inside the building by purlins and mid rails at frequent spacing. Polyurethane is a thermo-setting plastic which, while it burns, also forms a char layer. The end-use requirements for external cladding panels make it important that they are assembled with weather-tight joints and this may also help to improve fire performance. External cladding may be the subject of an arson attack, so adequate precautions need to be taken to minimise this risk.

Typical causes of ignition in reported fires in external claddings are:

- *arson;*
- *electrical/gas;*
- *rubbish/smoking.*

Appendix G: Fire safety information

1. Regulation 16B requires that, where building work involves the erection or extension of a relevant building, or a relevant change of use of a building, fire safety information shall be given to the responsible person at the completion of the project or when the building or extension is first occupied.

It is recommended that, for existing buildings, the building owner ensures as far as possible that full details, as defined in Appendix G, are made available to the insurer and fire authority.

- "fire safety information" means information relating to the design and construction of the building or extension, and the services, fittings and equipment provided in or in connection with the building or extension which will assist the responsible person to operate and maintain the building or extension with reasonable safety;
- a "relevant building" is a building to which the Regulatory Reform (Fire Safety) Order 2005 applies[1], or will apply after the completion of building work;
- a "relevant change of use" is a material change of use where, after the change of use takes place, the Regulatory Reform (Fire Safety) Order 2005 will apply, or continue to apply, to the building; and
- "responsible person" has the meaning given in article 3 of the Regulatory Reform (Fire Safety) Order 2005.

This Appendix is only intended as a guide as to the kind of information that should be provided. For clarity the guidance is given in terms of simple and complex buildings, however the level of detail required will vary from building to building and should be considered on a case by case basis.

Simple buildings

2. For most buildings basic information on the location of fire protection measures may be all that is necessary. An as-built plan of the building should be provided showing:

a. escape routes;

b. compartmentation and separation (i.e. location of fire separating elements, including cavity barriers in walk-in spaces);

c. fire doors, self-closing fire doors and other doors equipped with relevant hardware (e.g. panic locks);

d. locations of fire and/or smoke detector heads, alarm call-points, detection/alarm control boxes, alarm sounders, fire safety signage, emergency lighting, fire extinguishers, dry or wet risers and other fire fighting equipment and location of hydrants outside the building;

e. any sprinkler system(s), including isolating valves and control equipment;

f. any smoke-control system(s) (or ventilation system with a smoke-control function), including mode of operation and control systems;

g. any high-risk areas (e.g. heating machinery);

h. specifications of any fire safety equipment provided, in particular any routine maintenance schedules; and

i. any assumptions in the design of the fire safety arrangements regarding the management of the building.

j. Any provision incorporated into the building to facilitate the evacuation of Disabled people. This information can then be used when designing suitable Personal Emergency Escape Plans.

Complex buildings

3. For more complex buildings a more detailed record of the fire safety strategy and procedures for operating and maintaining any fire protection measures of the building will be necessary. Further guidance is available in BS5588-12:2004 *Fire precautions in the design, construction and use of buildings: Managing fire safety* (Annex A Fire Safety Manual.)

These records should include:

a. The fire safety strategy, including all assumptions in the design of the fire safety systems (such as fire load). Any risk assessments or risk analysis.

b. All assumptions in the design of the fire safety arrangements regarding the management of the building.

c. Escape routes, escape strategy (e.g. simultaneous or phased) and muster points.

d. Details of all passive fire safety measures, including compartmentation (i.e. location of fire separating elements), cavity barriers, fire doors, self-closing fire doors and other doors equipped with relevant hardware (e.g. electronic security locks), duct dampers and fire shutters.

[1] S.I. 2005/1541; *see* article 6.

e. Fire detector heads, smoke detector heads, alarm call-points, detection/alarm control boxes, alarm sounders, emergency communications systems, CCTV, fire safety signage, emergency lighting, fire extinguishers, dry or wet risers and other fire fighting equipment, other interior facilities for the fire and rescue service, emergency control rooms, location of hydrants outside the building, other exterior facilities for the fire and rescue service.

f. Details of all active fire safety measures, including:

- Sprinkler system(s) design, including isolating valves and control equipment; and
- Smoke-control system(s) (or HVAC system with a smoke-control function) design, including mode of operation and control systems.

g. Any high-risk areas (e.g. heating machinery) and particular hazards,

h. As-built plans of the building showing the locations of the above.

i. Specifications of any fire safety equipment provided, including operational details, operators manuals, software, system zoning and routine inspection, testing and maintenance schedules. Records of any acceptance or commissioning tests.

j. Any provision incorporated into the building to facilitate the evacuation of disabled people.

k. Any other details appropriate for the specific building.

Appendix H: Standards and other publications referred to

Standards

BS EN 54-11:2001
Fire detection and fire alarm systems. Manual call points

BS EN 81-72:2003
Safety rules for the construction and installation of lifts. Particular applications for passenger and goods passenger lifts. Firefighters lifts

BS EN 81-2:1998
Safety rules for the construction and installation of lifts. Hydraulic lifts

BS EN 81-1:1998
Safety rules for the construction and installation of lifts. Electric lifts

DD 252:2002
Components for residential sprinkler systems. Specification and test methods for residential sprinklers

BS EN ISO 306:2004
Plastics. Thermoplastic materials. Determination of Vicat softening temperature (VST)

BS 476-24:1987
Fire tests on building materials and structures. Method for determination of the fire resistance of ventilation ducts

BS 476-8:1972
Fire tests on building materials and structures. Test methods and criteria for the fire resistance of elements of building construction (withdrawn)

BS 476-23:1987
Fire tests on building materials and structures. Methods for determination of the contribution of components to the fire resistance of a structure

BS 476-22:1987
Fire tests on building materials and structures. Methods for determination of the fire resistance of non-loadbearing elements of construction

BS 476-4:1970
Fire tests on building materials and structures. Non-combustibility test for materials

BS 476-21:1987
Fire tests on building materials and structures. Methods for determination of the fire resistance of loadbearing elements of construction

BS 476-20:1987
Fire tests on building materials and structures. Method for determination of the fire resistance of elements of construction (general principles)

BS 476-11:1982
Fire tests on building materials and structures. Method for assessing the heat emission from building materials

BS 476-6:1989
Fire tests on building materials and structures. Method of test for fire propagation for products

BS 476-7:1997
Fire tests on building materials and structures. Method of test to determine the classification of the surface spread of flame of products

BS 476-3:2004
Fire tests on building materials and structures. Classification and method of test for external fire exposure to roofs

BS EN 771-1:2003
Specification for masonry units. Clay masonry units

BS EN 1634-2:xxxx
Fire resistance tests for door and shutter assemblies, Part 2 – Fire door hardware

BS EN 1125:1997
Building hardware. Panic exit devices operated by a horizontal bar. Requirements and test methods

BS EN 1155:1997
Building hardware. Electrically powered hold-open devices for swing doors. Requirements and test methods

BS EN ISO 1182:2002
Reaction to fire tests for building products. Non-combustibility test

ENV 1187:2002, test 4
Test methods for external fire exposure to roofs

BS EN 1363-1:1999
Fire resistance tests. General requirements

BS EN 1364-1:1999
Fire resistance tests for non-loadbearing elements. Walls

BS EN 1364-2:1999
Fire resistance tests for non-loadbearing elements. Ceilings

BS EN 1364-3:2006
Fire resistance tests for non-loadbearing elements. Curtain walling. Full configuration (complete assembly)

BS EN 1365-1:1999
Fire resistance tests for loadbearing elements. Walls

BS EN 1365-2:2000
Fire resistance tests for loadbearing elements. Floors and roofs

BS EN 1365-3:2000
Fire resistance tests for loadbearing elements. Beams

BS EN 1365-4:1999
Fire resistance tests for loadbearing elements. Columns

BS EN 1366-1:1999
Fire resistance tests for service installations. Fire resistance tests for service installations. Ducts

BS EN 1366-2:1999
Fire resistance tests for service installations. Fire dampers

BS EN 1366-3:2004
Fire resistance tests for service installations. Penetration seals

BS EN 1366-4:2006
Fire resistance tests for service installations. Linear joint seals

BS EN 1366-5:2003
Fire resistance tests for service installations. Service ducts and shafts

BS EN 1366-6:2004
Fire resistance tests for service installations. Raised access and hollow core floors

BS EN 1634-1:2000
Fire resistance tests for door and shutter assemblies. Fire doors and shutters

BS EN 1634-3:2001
Fire resistance tests for door and shutter assemblies. Smoke control doors and shutters

BS EN ISO 1716:2002
Reaction to fire tests for building products. Determination of the heat of combustion

BS 2782-0:2004
Methods of testing. Plastics. Introduction

BS 3251:1976
Specification. Indicator plates for fire hydrants and emergency water supplies

BS 4514:2001
Unplasticized PVC soil and ventilating pipes of 82.4mm minimum mean outside diameter, and fittings and accessories of 82.4mm and of other sizes. Specification

BS 5255:1989
Specification for thermoplastics waste pipe and fittings

BS 5266-1:2005
Emergency lighting.Code of practice for the emergency lighting of premises

BS 5306-2:1990
Fire extinguishing installations and equipment on premises. Specification for sprinkler systems

BS 5395-2:1984
Stairs, ladders and walkways. Code of practice for the design of helical and spiral stairs

BS 5438:1989
Methods of test for flammability of textile fabrics when subjected to a small igniting flame applied to the face or bottom edge of vertically oriented specimens

BS 5446-1:2000
Fire detection and fire alarm devices for dwellings. Specification for smoke alarms

BS 5446-2:2003
Fire detection and fire alarm devices for dwellings. Specification for heat alarms

BS 5499-1:2002
Graphical symbols and signs. Safety signs, including fire safety signs. Specification for geometric shapes, colours and layout

BS 5499-1:2002
Graphical symbols and signs. Safety signs, including fire safety signs. Specification for geometric shapes, colours and layout

BS 5588-1:1990
Fire precautions in the design, construction and use of buildings. Code of practice for residential buildings

BS 5588-5:2004
Fire precautions in the design, construction and use of buildings. Access and facilities for fire-fighting

BS 5588-6:1991
Fire precautions in the design, construction and use of buildings. Code of practice for places of assembly

BS 5588-7:1997
Fire precautions in the design, construction and use of buildings. Code of practice for the incorporation of atria in buildings

BS 5588-8:1999
Fire precautions in the design, construction and use of buildings. Code of practice for means of escape for disabled people

BS 5588-9:1999
Fire precautions in the design, construction and use of buildings. Code of practice for ventilation and air conditioning ductwork

BS 5588-10:1991
Fire precautions in the design, construction and use of buildings. Code of practice for shopping complexes

BS 5588-11:1997
Fire precautions in the design, construction and use of buildings. Code of practice for shops, offices, industrial, storage and other similar buildings

BS 5588-12:2004
Fire precautions in the design, construction and use of buildings. Managing fire safety

BS 5720:1979
Code of practice for mechanical ventilation and air conditioning in buildings

BS 5839-8:1998
Fire detection and fire alarm systems for buildings. Code of practice for the design, installation, commissioning, and maintenance of voice alarm systems

BS 5839-1:2002
Fire detection and fire alarm systems for buildings. Code of practice for system design, installation, commissioning and maintenance

BS 5839-1:2002 + A2:2008
Fire detection and fire alarm systems for buildings. Code of Practice for system design, installation, commissioning and maintenance

BS 5839-6:2004
Fire detection and fire alarm systems for buildings. Code of practice for the design, installation and maintenance of fire detection and fire alarm systems in dwellings

BS 5839-9:2003
Fire detection and alarm systems for buildings. Code of practice for the design, installation, commissioning and maintenance of emergency voice communication systems

BS 5867-2:1980
Specification for fabrics for curtains and drapes. Flammability requirements

BS 5950-8:2003
Structural use of steelwork in building. Code of practice for fire resistant design

BS 6336:1998
Guide to the development of fire tests, the presentation of test data and the role of tests in hazard assessment

BS 7157:1989
Method of test for ignitability of fabrics used in the construction of large tented structures

BS 7273-2:1992
Code of practice for the operation of fire protection measures. Mechanical actuation of gaseous total flooding and local application extinguishing systems

BS 7273-3:2000
Code of practice for the operation of fire protection measures. Electrical actuation of pre-action sprinkler systems

BS 7273-1:2006
Code of practice for the operation of fire protection measures. Electrical actuation of gaseous total flooding extinguishing systems

BS 7346-6:2005
Components for smoke and heat control systems. Specifications for cable systems

BS 7346-7:2006
Components for smoke and heat control systems. Code of practice on functional recommendations and calculation methods for smoke and heat control systems for covered car parks

BS 7974:2001
Application of fire safety engineering principles to the design of buildings. Code of practice

BS 8414-2:2005
Fire performance of external cladding systems. Test method for non-loadbearing external cladding systems fixed to and supported by a structural steel frame

BS 8414-1:2002
Fire performance of external cladding systems. Test methods for non-loadbearing external cladding systems applied to the face of a building

BS 8214:1990
Code of practice for fire door assemblies with non-metallic leaves

BS 8214:1990

Code of practice for fire door assemblies with non-metallic leaves

BS 8313:1997
Code of practice for accommodation of building services in ducts

BS 9251:2005
Sprinkler systems for residential and domestic occupancies. Code of practice

BS 9990:2006
Code of practice for non-automatic fire-fighting systems in buildings

BS ISO 10294-2:1999
Fire-resistance tests. Fire dampers for air distribution systems. Classification, criteria and field of application of test results

BS ISO 10294-5:2005
Fire-resistance tests. Fire dampers for air distribution systems. Intumescent fire dampers

BS EN ISO 11925-2:2002
Reaction to fire tests. Ignitability of building products subjected to direct impingement of flame. Single-flame source test

BS EN 12101-3:2002
Smoke and heat control systems. Specification for powered smoke and heat exhaust ventilators

BS EN 12101-3:2002
Smoke and heat control systems. Specification for powered smoke and heat exhaust ventilators

BS EN 12101-6:2005
Smoke and heat control systems. Specification for pressure differential systems. Kits

BS EN 12845:2004
Fixed firefighting systems. Automatic sprinkler systems. Design, installation and maintenance

BS EN 13238:2001
Reaction to fire tests for building products. Conditioning procedures and general rules for selection of substrates

BS EN 13501-1:2002
Fire classification of construction products and building elements. Classification using test data from reaction to fire tests

BS EN 13501-2:2003
Fire classification of construction products and building elements. Classification using data from fire resistance tests, excluding ventilation services

BS EN 13501-3:2005
Fire classification of construction products and building elements. Classification using data from fire resistance tests on products and elements used in building service installations: fire resisting ducts and fire dampers

BS EN 13501-4:xxxx
Fire classification of construction products and building elements, Part 4 – Classification using data from fire resistance tests on smoke control systems

BS EN 13501-5:2005
Fire classification of construction products and building elements. Classification using data from external fire exposure to roof tests

BS EN 13823:2002
Reaction to fire tests for building products. Building products excluding floorings exposed to the thermal attack by a single burning item

BS EN 50200:2006
Method of test for resistance to fire of unprotected small cables for use in emergency circuits

BS EN 50289-4-11:2002
Communication cables. Specifications for test methods. Environmental test methods. A horizontal integrated fire test method

Publications

Legislation

Disability Discrimination Act 1995

Education Act 1996

Pipelines Safety Regulations 1996, SI 1996 No 825 and the Gas Safety (Installation and Use) Regulations 1998 SI 1998 No 2451

Electromagnetic Compatibility Regulations 1992 (SI 1992 No 2372)

Electromagnetic Compatibility (Amendment) Regulations 1994 (SI 1994 No 3080)

Electrical Equipment (Safety) Regulations 1994 (SI 1994 No 3260)

Commission Decision 2000/553/EC of 6th September 2000 implementing Council Directive 89/106/EEC

(European tests) Commission Decision 2000/367/EC of 3rd May 2000 implementing Council Directive 89/106/EEC

Commission Decision 2001/671/EC of 21 August 2001 implementing Council Directive 89/106/EC as regards the classification of the external fire performance of roofs and roof coverings

Commission Decision 2005/823/EC of 22 November 2005 amending Decision 2001/671/EC regarding the classification of the external fire performance of roofs and roof coverings

Commission Decision 2000/147/EC of 8th February 2000 implementing Council Directive 89/106/EEC

Commission Decision 2000/367/EC of 3rd May 2000 implementing Council Directive 89/106/EEC

Commission Decision 96/603/EC of 4th October 1996

94/61 1/EC implementing Article 20 of the Council Directive 89/106/EEC on construction products

Construction Products Regulations 1991 (SI 1991 No 1620)

Construction Product (Amendment) Regulations 1994 (SI 1994 No 3051)

The Workplace (Health, Safety and Welfare) Regulations 1992

Health and Safety (Safety signs and signals) Regulations 1996

Association for Specialist Fire Protection (ASFP)

ASFP Red book – *Fire stopping and penetration seals for the construction industry* 2nd Edition ISBN: 1 87040 923 X

ASFP Yellow book – *Fire protection for structural steel in buildings* 4th Edition ISBN: 1 87040 925 6

ASFP Grey book – *Fire and smoke resisting dampers* ISBN: 1 87040 924 8

ASFP Blue book – *Fire resisting ductwork* 2nd Edition ISBN: 1 87040 926 4

www.asfp.org.uk

The British Automatic Sprinkler Association (BAFSA)

Sprinklers for Safety: Use and Benefits of Incorporating Sprinklers in Buildings and Structures, (2006) ISBN: 0 95526 280 1

www.bafsa.org.uk

Building Research Establishment Limited (BRE)

BRE Digest 208 *Increasing the fire resistance of existing timber floors* 1988 ISBN: 978 1 86081 359 7

BRE report (BR 368) *Design methodologies for smoke and heat exhaust ventilation* 1999 ISBN: 978 1 86081 289 7

BRE report (BR 274) *Fire safety of PTFE-based materials used in buildings* 1994 ISBN: 978 1 86081 653 6

BRE report (BR 135) *Fire performance of external thermal insulation for walls of multi-storey buildings* 2003 ISBN: 978 1 86081 622 2

BRE report (BR 187) *External fire spread: Building separation and boundary distances* 1991 ISBN: 978 1 86081 465 5

BRE report (BR128) *Guidelines for the construction of fire resisting structural elements* 1988 ISBN: 0 85125 293 1

BRE 454 *Multi-storey timber frame buildings – a design guide* 2003 ISBN: 1 86081 605 3

LPS 1056: Issue 6.1: *Test and evaluation requirements for the LPCB approval and listing of fire doorsets, lift landing doors and shutters,* 2005

LPS 1181-1: Issue 1.1: *Series of fire growth tests for LPCB approval and listing of construction product systems: Requirements and tests for built-up cladding and sandwich panel systems for use as the external envelope of buildings,* 2005

LPS 1181-2: Issue 2.0: *Series of fire growth tests for LPCB approval and listing of construction product systems: Requirements and tests for sandwich panels and built-up systems for use as internal constructions in buildings,* 2005

LPS 1208: Issue 2.1: *LPCB fire resistance requirements for elements of construction used to provide compartmentation,* 2005

LPS 1263: Issue 1.1: *Requirements for the LPCB approval and listing of the fire performance of kitchen extract systems,* 2005

www.bre.co.uk

Builders Hardware Industry Federation and Guild of Architectural Ironmongers

Hardware for Fire and Escape Doors 2006 ISBN: 0 95216 422 1

www.firecode.org.uk

Department for Communities and Local Government

Regulatory Reform (Fire Safety) Order 2005 ISBN: 0 11072 945 5

Fire safety in adult placements: a code of practice

www.communities.gov.uk

Department for Education and Skills

Building Bulletin (BB) 100

www.dfes.gov.uk

Department of Health

HTM 05 – 02 Guidance in support of functional provisions for healthcare premises

www.dh.gov.uk

Door and Shutter Manufacturers' Association (DSMA)

Code of practice for fire-resisting metal doorsets 1999

www.dhfonline.org.uk

Environment Agency

Pollution Prevention Guidelines (PPG18) *Managing Fire Water and Major Spillages*

www.environment-agency.gov.uk

Football Licensing Authority

Concourses ISBN: 0 95462 932 9

www.flaweb.org.uk/home.php

Fire Protection Association (FPA)

FPA Design guide for the fire protection of buildings: Essential principles, Fire Protection Association for InFiReS, 2007

FPA Design guide for the fire protection of buildings: Core document: Compartmentation, Fire Protection Association for InFiReS, 2008

FPA Design guide for the fire protection of buildings: Protected zone, Fire Protection Association for InFiReS, 2004

FPA Design guide: The fire protection of buildings: Core document: Protection of openings and service penetrations from fire, Fire Protection Association for InFiReS, 2005

FPA Design guide for the fire protection of buildings: Warehouses and storage buildings 1: Design principles, Fire Protection Association for InFiReS, 2008

LPC Rules for automatic sprinkler installations incorporating BS EN 12845, Fire Protection Association, 2007

The prevention and control of arson, 2nd edition, Fire Protection Association, 2008

Fire safety and waste materials, Fire Protection Association, 2003

www.thefpa.co.uk

Glass and Glazing Federation (GGF)

A guide to best practice in the specification and use of fire-resistant glazed systems

www.ggf.org.uk

Health and Safety Executive (HSE)

Workplace health, safety and welfare, The Workplace (Health, Safety and Welfare) Regulations 1992, Approved Code of Practice and Guidance; The Health and Safety Commission, L24; published by HMSO 1992; ISBN: 0 11886 333 9

www.hse.gov.uk

International Association of Cold Storage Contractors (IACSC)

Design, construction, specification and fire management of insulated envelopes for temperature controlled environments 1999

www.iarw.org/iacsc/european_division

Intumescent Fire Seals Association

The IFSA Code, *Sealing apertures and service penetrations to maintain fire resistance*, 1999

www.ifsa.org.uk

For Loss Prevention Standards, please see Building Research Establishment Limited (BRE).

National Fire Protection Association (USA)

NFPA 262: *Standard method of test for flame travel and smoke of wires and cables for use in air-handling spaces*, 2007

Passive Fire Protection Federation

Ensuring best practice for passive fire protection in buildings ISBN: 1 87040 919 1

www.pfpf.org

Steel Construction Institute (SCI)

SCI P197 Designing for structural fire safety: A handbook for architects and engineers 1999 ISBN: 1 85942 074 5

SCI Publication 288 Fire safe design: A new approach to multi-storey steel-framed buildings (Second Edition) 2000 ISBN: 1 85942 169 5

SCI Publication P313 Single storey steel framed buildings in fire boundary conditions 2002 ISBN: 1 85942 135 0

www.steel-sci.org

Timber Research and Development Associations (TRADA)

Timber Fire-Resisting Doorsets: maintaining performance under the new European test standard ISBN: 1 90051 035 9

www.trada.co.uk

Appendix I: Main changes in the 2006 edition

This is the second impression of the 2006 Edition of Approved Document B which incorporates the typographical and proofing corrections issued in April 2007.

This edition of Approved Document B, Fire safety, replaces the 2000 edition. The main changes are:

General

a. **Approved Document B:** The Approved Document has been split into two volumes. Volume 1 deals with dwellinghouses, Volume 2 deals with buildings other than dwellinghouses.

b. **Fire Safety Information:** A new Regulation (16B) has been introduced to ensure that sufficient information is recorded to assist the eventual owner/occupier/employer to meet their statutory duties under the Regulatory Reform (Fire Safety) Order 2005.

Introduction

c. **Management of Premises:** New guidance is given on the need to ensure that management regimes are realistic.

d. **Certification Schemes:** Suitable schemes may be accepted by Building Control Bodies as evidence of compliance.

e. **Residential Sprinklers:** The use of sprinkler systems in accordance with BS 9251:2005 is recognised.

f. **Alternative Approaches**

  i. HTM 05 "Firecode" should be used for used for the design of hospitals and similar health care premises.

  ii. BB100 should be used for the design of schools.

B1

g. **Fire Alarms in Flats:** The guidance on smoke alarms in flats has been amended such that alarms should be installed in accordance with BS 5839-6:2004.

Simple guidance has been retained, in the form of a commentary on this standard, so that most users of the Approved Document will necessarily not need to obtain a copy of the standard.

  i. All smoke alarms should have a standby power supply.

h. **Fire Alarms:** The guidance for buildings other than dwellings has been updated to take account of the 2002 edition of BS 5839-1.

i. **Means of escape:**

  i. Locks and child resistant safety stays may be provided on escape windows.

  ii. New guidance has been provided on the provision of galleries and inner inner rooms.

  iii. Additional options of providing sprinkler protection and/or a protected stairway instead

of alternative escape routes has been included for flats with more than one storey.

  iv. Guidance on the use of air circulation systems in flats with protected entrance halls or stairways is given.

  v. The provisions for smoke control in the common areas of flats have been changed.

  vi. Guidance on means of escape in buildings with open spatial planning has been included.

  vii. A method has been provided for calculating acceptable final exit widths for merging escape routes at ground floor level.

  viii. Guidance on the provision of cavity barriers associated with subdivided corridors has been moved to Section 3 and clarified.

  ix. Guidance applicable to small premises, previously in BS 5588-11, has been incorporated into the text.

  x. New guidance on the design of residential care homes has been given including the use of sprinklers and/or free swing door closing devices. Greater flexibility is also given where sprinkler systems are provided.

  xi. Guidance on means of escape for disabled people has been incorporated in the general guidance on means of escape.

  xii. In tall building with phased evacuation consideration needs to be given to the interaction of firefighters with people attempting to evacuate the building.

  xiii. More detailed guidance has been provided on the protection of ventilation systems.

B3

j. **Compartment Walls:** The predicted deflection of a floor, in the event of a fire, should be accommodated in the design of compartment walls.

k. **Sprinkler protection in flats:** Sprinkler systems should be provided in blocks of flats exceeding 30m in height.

l. **Warehouses:** A maximum compartment size has been introduced for unsprinklered single-storey warehouse buildings.

m. **Concealed Spaces:** This section has been completely restructured. Table 13 (AD B 2000) has now been incorporated into the text to reduce confusion.

   i. **Openings:** Window and door frames should only be regarded as cavity barriers if they are constructed of steel or timber of an appropriate thickness.

   ii. **Under Floor Voids:** Extensive cavities in floor voids should be subdivided with cavity barriers.

n. **Fire dampers:** Guidance on the specification and installation of fire dampers has been provided.

o. **Car Parks:** Non combustible materials should be used in the construction of a car park for it to be regarded as 'open sided' for the purposes of establishing the necessary period of fire resistance. Other car parks should achieve the standard period of fire resistance.

B4

p. **Notional Boundaries:** Space separation should be considered where more than one building is on the same site but operated by different 'organisations'.

q. **Roof Coverings:** The guidance on roof coverings incorporates the new European system of classification set out in BS EN 13501-5:2005.

B5

r. **Private Hydrants:** A building with a compartment of 280m^2 or more, constructed more than 100m from a highway, should be provided with suitable fire hydrants.

s. **Vehicle Access:** There should be access for a pump appliance to within 45m of all points within a dwelling or a suitable fire main should be provided

t. **Firefighting Shafts:** Assembly buildings with a floor exceeding 900m^2, with a floor over 7.5m above ground level, should be provided with firefighting shafts.

u. **Hose distances:** In unsprinklered buildings every part of every storey over 18m in height should be within 45m of a fire main outlet.

Appendix B

v. **Self-Closing Devices:** Fire doors within flats need not be provided with self-closing devices.

Appendix C

w. **Occupant capacity:** The floor space factors table has been updated and moved to this Appendix.

x. **Door width:** The method of measurement for door width has been changed to align with Approved Document M.

y. **Smoke ventilators:** Guidance is given on the measurement of free area for smoke ventilators.

Appendix G

z. **Fire Safety Information:** This new Appendix provides guidance on the new requirement for fire safety information to be recorded and passed on to the 'responsible' person.

Appendix J: Insurer requirements for the implementation of fire safety engineering solutions

Contents

Section 1

Section 2

Section 1 Use of guidance

1.1 Background

1.1.1 What is fire safety engineering?

The Institution of Fire Engineers (IFE) defines fire safety engineering as 'the application of scientific and engineering principles, rules and expert judgement, based on an understanding of the phenomena and effects of fire and of the reaction and behaviour of people to fire, to protect people, property and the environment from the destructive effects of fire.' For the purposes of this document, the term 'fire safety engineer' does not include those who simply apply prescriptive fire safety codes and standards to the design of buildings. Neither does it include those who undertake fire risk assessments of such premises using simple techniques and checklists.

Fire safety engineers use their knowledge of fire behaviour, reaction of materials to fire and human behaviour in a fire to develop unique fire safety solutions. This work often involves calculation and the use of engineering judgement.

Fire safety engineering is being used more and more frequently in the UK as an alternative to the application of prescriptive fire safety codes and standards. It may be the only practical way to achieve a satisfactory level of fire safety in some large and complex buildings, and in buildings containing different types of occupancy. Additionally, it offers the designer more freedom to create innovative buildings, and, because fire safety engineering removes the possibility of unquantified 'over-design' related to the prescriptive approach, a fire safety solution may offer a more cost-effective solution.

When developing fire safety engineered solutions for buildings, it is common practice for the fire safety engineer to be instructed to concentrate solely on life safety and evacuation because these are the elements mandated by the building regulations process. Although life safety is of utmost importance, a solution which focuses exclusively on life safety can have a detrimental effect on property and business protection when compared with a prescriptive codes solution. This is of concern to the insurance industry and it is the purpose of this document to address that concern.

1.1.2 About this document

Although this document has been written on behalf of insurers, the technical content is intended for use by fire safety engineers and everyone who has involvement in the development and delivery of a fire engineering solution. The format and framework that it follows is based on the British Standard for fire safety engineering in buildings, BS 7974: 2001: *Application of fire safety engineering principles to the design of buildings. Code of practice* (Ref. 1; see also 'Note regarding BS 7974' below). The techniques described will easily be recognised by competent fire safety engineers.

Throughout this document, the importance of contact and consultation with the insurer is stressed. This is because this document is not a 'national standard'. All insurers have their own views on fire safety engineering, its risks and the appropriate ways of addressing those risks. This document simply describes the broad principles which the fire safety engineer should follow. Where detail is given, it is only by way of example.

It should be highlighted that the insurer of a project during its design and build phase will often **not** be the organisation that will finally end up insuring the building on occupation. If the final insurer has not been identified at the design stage of the project, then the fire safety engineer has little choice but to follow the broad principles described here. However, it must be highlighted that the specific expectations of individual insurers will differ, and standards which are higher (or lower) than those quoted may be called for.

If both the design and build phase insurer **and** the final insurer can be identified, then the whole project should be signed off by both. In the event of a conflict of interest between these two insurers, priority should be with the final insurer, but this is a matter which must be resolved by the insurers themselves, not the fire safety engineer.

Consultation will usually involve non-technical discussions between the insurer, the fire safety engineer, the designer and the end user regarding the insurer's specific expectations in terms of property protection, the avoidance of business interruption, fire safety management and so on. If the fire safety engineer wants detailed technical guidance, or if the insurer wishes to understand a particular point of fire engineering practice, this document will not provide an explanation.

Insurers can obtain fully independent technical guidance on matters of fire safety engineering from the Fire Engineering Services group of the Technical Division of the FPA.

Fire Protection Association
Technical Division
London Road
Moreton in Marsh
Gloucestershire
GL56 0RH
Tel: 01608 812 500
Email: fse@thefpa.co.uk

Note regarding BS 7974

The following statement is taken from the Foreword to BS 7974:

'BS 7974 is intended to provide a framework for the application of fire safety engineering principles to the design of buildings. It is supported by the PD 7974 series of Published Documents that contain guidance and information on how to undertake detailed analysis of specific aspects of fire safety engineering in buildings. PD 7974-0 to 7 provide a summary of the state of the art and it is intended that they will be updated as new theories, calculation methods and/or data become available. PD 7974 is structured as follows:

- *Part 0: Guide to the design framework and fire safety engineering procedures;*
- *Part 1: (Sub-system 1) Initiation and development of fire within the enclosure of origin;*
- *Part 2: (Sub-system 2) Spread of smoke and toxic gases within and beyond the enclosure of origin;*
- *Part 3: (Sub-system 3) Structural response and fire spread beyond the enclosure of origin;*
- *Part 4: (Sub-system 4) Detection of fire and activation of fire protection systems;*
- *Part 5: (Sub-system 5) Fire service intervention;*
- *Part 6: (Sub-system 6) Evacuation;*
- *Part 7: Probabilistic fire risk assessment.*

This code of practice [BS 7974] can be used to identify and define one or more fire safety design issues to be addressed using fire safety engineering. The appropriate part(s) of PD 7974 can then be used to set specific acceptance criteria and undertake detailed analysis.'

1.2 Foreword

The *FPA Design guide for the fire protection of buildings: Essential principles* (Ref. 2) document states that fire safety engineering may be used as an alternative to following the guidance contained in the *FPA Design guide*, but:

- it is important that early consultation with insurers during the building design phase is carried out. This should ensure that the most effective fire protection measures appropriate to the specific property, end use application and business protection needs are suitably satisfied;
- an appropriate fire safety engineering standard must be followed in full. BS 7974 is recommended, although *CIBSE Guide E: Fire engineering* (Ref. 3) is also acceptable;
- the essential principles of the *FPA Design guide* must be achieved to the extent that they would be if the *Design guide* was followed in full (when assessed as a comparative approach); and
- the objectives of the fire safety engineering design must include the:
  - preservation of the structure and fabric of the building;
  - preservation of the building contents;
  - protection of ongoing business viability; and
  - maintenance of the corporate image.

This document explains a series of requirements placed upon fire safety engineering solutions which must be followed if the essential principles of the *Design guide* are to be met in such buildings.

1.3 Application

The requirements contained in this document are applicable to any new building, and to the alteration of any existing building, in which the fire safety design does not follow the solutions contained in approved guidance documents – including the *FPA Design guide* – and where fire safety engineering is to be used to provide an alternative approach to fire safety.

Where the fire safety of a building has been designed using the principles of fire safety engineering and this document, the building may be limited in terms of its flexibility of use. Therefore, the suitability of an existing fire engineered building may need to be re-assessed using this document when it undergoes a change such as a change of use, a change of occupier, a material alteration or a change in the nature of materials stored or processes carried out.

In general, fire safety engineering design only considers the fire safety of a building once it has been occupied. However, buildings are at their most vulnerable to fire during their construction phase. Because of their reliance on active fire safety measures, such as fire detection and/or suppression, this can be particularly true in the case of buildings where fire safety engineering has been employed. As a result, the requirements contained in this document include fire safety considerations which must be followed during construction or alteration.

1.4 Terms and definitions

1.4.1 Acceptance criteria

Those criteria, including performance requirements and essential conditions, which must be met before the fire engineering design/strategy is accepted.

In practice, following quantified analysis, the results need to be compared with the acceptance criteria identified during the QDR (see below). Three basic approaches can be considered by the QDR team when setting acceptance criteria:

- comparative;
- deterministic;
- probabilistic.

These are defined in PD 7974-0.

1.4.2 Fire safety engineering

The application of scientific and engineering principles, rules and expert judgement, based on an understanding of the phenomena and effects of fire and of the reaction and behaviour of people to fire, to protect people, property and the environment from the destructive effects of fire (definition from the Institution of Fire Engineers).

1.4.3 Fire load

The sum of the calorific energies which could be released by the complete combustion of all the combustible materials in a space, including the facings of the walls, partitions, floors and ceilings.

1.4.4 Fire load density

The fire load divided by the floor area.

1.4.5 Qualitative design review (QDR)

A meeting at which the scope and objectives of the fire safety design are defined, performance criteria established and one or more potential design solutions proposed. Key information is also gathered to enable evaluation of the design solutions in the quantitative analysis.

Section 2 Essential principles and fire engineering requirements

The objectives of this document are based upon the essential principles of the *FPA Design guide*. These objectives may be met by following a series of principles.

2.1 Essential principles

While the detailed recommendations of the *FPA Design guide* are not mandatory, the essential principles shall be regarded as being of vital importance and need to be considered in detail by building designers, fire safety engineers, consultants, occupiers or building owners.

It is essential that insurers are consulted at the earliest possible stage of design, so that factors that influence the ability of the building owner to secure insurance at the most attractive terms can be identified.

Reaction in the event of fire	
PRINCIPLE 1	The building shall be constructed in such a manner that if a fire starts, the extent of fire and smoke damage will be minimised and confined as close to the source of fire outbreak as is practical/feasible.
PRINCIPLE 2	With the exception of joinery products, the building shall be constructed from building materials/products that will not make a significant contribution to the early stages of a fire or contribute to the spread of fire.
PRINCIPLE 3	Suitable measures will be taken for the prevention of premature structural collapse and excessive deflection.
PRINCIPLE 4	Consideration should be given at the design stage regarding potential damage from firefighting water and to ensure as far as practical that the effect on the environment of the fire effluent will be minimised.
Workmanship	
PRINCIPLE 5	As a minimum, all fire protection products shall be third-party certified to an appropriate product or performance-based standard (attestation level 1 for CE marking).
PRINCIPLE 6	All fire protection products/systems shall be installed by adequately trained specialist installers.
Response to fire	
PRINCIPLE 7	The building shall be fitted with an appropriate automatic fire alarm system.
PRINCIPLE 8	The fire protection systems shall be regularly maintained so that they are able to perform their intended function throughout the life of the building.
Fire prevention	
PRINCIPLE 9	There shall be adequate provision to prevent an arson attack.
PRINCIPLE 10	The building shall be so constructed that fire cannot spread into the premises from an adjoining building or other external fire source.
Fire safety management	
PRINCIPLE 11	The building owner shall ensure an adequate standard of fire safety management throughout the life of the building.
PRINCIPLE 12	Any fuel-burning appliance and services or electrical appliance and services shall be designed, constructed and installed in a manner that reduces their potential as an accidental source of ignition.

2.2 Summary of requirements

This document defines a series of requirements placed upon fire safety engineering solutions which must be followed if the essential principles of the *Design guide* are to be met in such buildings.

REQUIREMENT 1	To meet the requirements of the insurance industry, wherever practicable, the insurer must be invited to join the QDR team. Buildings are particularly vulnerable to fire during the construction phase, so the invitation must be extended to the insurer of the building during its construction phase or to the principal contractor's insurer.
REQUIREMENT 2	To meet the requirements of the insurance industry, the fire safety objectives of the QDR must include the matters detailed in this document to the extent determined by the agreed acceptance criteria.
REQUIREMENT 3	In order to minimise insured losses, fire suppression or containment (where used) must be initiated prior to critical damage to property or business taking place.
REQUIREMENT 4a	Table 9 of PD 7974-0 (Ref. 4) gives an example of property protection objectives which may be set as deterministic acceptance criteria. In this document, Table 1 presents objectives which must be met in order to satisfy insurer requirements.
REQUIREMENT 4b	If a probabilistic approach is used to set acceptance criteria, then the objectives which must be met in order to satisfy insurer requirements must be discussed with the insurer. Broad guidance is described elsewhere in this document.
REQUIREMENT 4c	The degree of structural protection, loss prevention and business continuity assurance provided by a fire safety engineering solution must be at least as high as that provided by a fire safety design for an equivalent building which fully complies with an appropriate code or standard.
REQUIREMENT 5	To enable a third party to establish that the fire safety engineering study has been carried out by a person with appropriate expertise, the name, qualifications and experience of the individual fire safety engineer(s) responsible for the study must be provided. In addition, the fire engineer(s) will follow a recognised set of fire engineering guidelines (such as BS 7974) from the beginning to the end of a project. The fire safety engineer(s) must provide a note of their current professional indemnity insurer, including the policy number and limit of indemnity.
REQUIREMENT 6	To meet the requirements of the insurance industry, the fire engineering solution must take a realistic view of the level of fire safety management which will be applied to the building during its foreseeable future, and must demonstrate that this level of management is sustainable. The fire safety engineer must demonstrate that every effort has been made to gather this information. Fire safety management requirements such as testing, servicing and maintenance (in particular those aspects upon which the fire safety engineering relies) must be described in the fire safety engineer's report.
REQUIREMENT 7	Fire safety management of construction sites is always important, but to meet the requirements of the insurance industry, where the fire safety design of a building is engineered, then a risk assessment of the implications of the fire engineering to the construction phase fire risk must be carried out and must be available to the insurer.
REQUIREMENT 8	With respect to the issues addressed by this document, the fire engineer's report must contain a non-technical explanation of how requirements 1 to 7 have been met. In the case of large or complex projects, it may be necessary to produce this information as a separate report for the insurer.

2.3 Stakeholders

2.3.1 Regulatory requirements

In order to gain approval for the construction of most buildings in England and Wales, their design must comply with the current Building Regulations (Ref. 5). In Scotland, designs must comply with the Building (Scotland) Regulations (Ref. 6), and in Northern Ireland, the Building Regulations (Northern Ireland) (Ref. 7).

In terms of fire safety, guidance on meeting the regulatory requirements described above is given in the following documents:

England and Wales:	Approved Document B to the Building Regulations 2000: *Fire Safety (Volume 2): Buildings other than Dwelling Houses* (Ref. 8)
Scotland:	*Scottish Building Standards Technical Handbook (Non-domestic)*, Section 2: Fire (Ref. 9)
Northern Ireland:	Technical Booklet E to the Building Regulations (Northern Ireland) 2000: *Fire Safety* (Ref. 10)

The primary emphasis of all these documents is the prevention of injury and death in the event of fire.

For example, *Technical Booklet E* (for Northern Ireland) says that:

'Building Regulations are intended to ensure that an adequate standard of life safety is provided in case of fire. The protection of property, including the building itself, may require additional measures and insurers will in general seek their own measures if they are to accept the insurance risk.'

It is therefore tempting to assume that a building constructed in compliance with such guidance would contain *no* measures to protect the structure of the building or its contents from fire. However, this assumption would be wrong.

The approach taken by the guidance documents is that life safety is delivered by a combination of methods. In general, these methods include reasonable means of escape from fire, reasonable means to prevent fire spread, reasonable measures to ensure structural stability in the early stages of fire and reasonable measures for fire service access (see Figure 1a).

Consequently, although the only regulatory requirement for the design of new buildings is that they must ensure the safety of occupants in the event of fire, a by-product of this requirement is that the structure of a building and its contents usually have a degree of protection from fire as well.

2.3.2 Insurer interest

During construction and occupation, buildings are usually insured. Historically, the risk which insurers have taken on has benefited from an element of structural fire protection which has been provided as a consequence of prescriptive life safety requirements. Premiums have always been calculated based on this assumption.

However, as has already been described, a performance-based approach can be employed to meet the same life safety objectives without providing structural fire resistance (see Figure 1b).

To ensure that the most effective fire safety measures appropriate to the specific property are applied, early consultation with the insurer and consideration of this document is essential. Failure to do so at an early stage could result in additional measures being required after construction, the use of the building being restricted or insurance premiums and/or deductibles being increased.

Figure 1a: Mandatory life safety standards deliver a level of life safety by a combination of means of escape and building protection

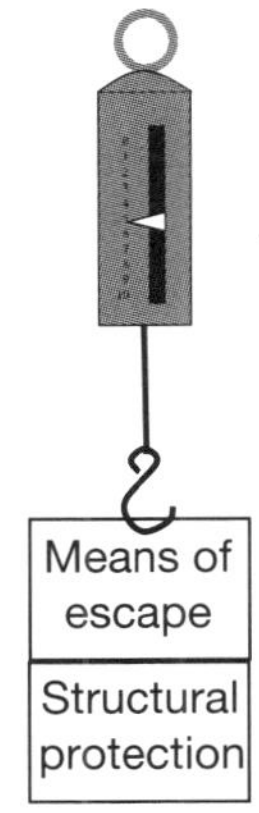

Figure 1b: Fire safety engineering can be used to deliver the same level of life safety, but dispensing with any building protection

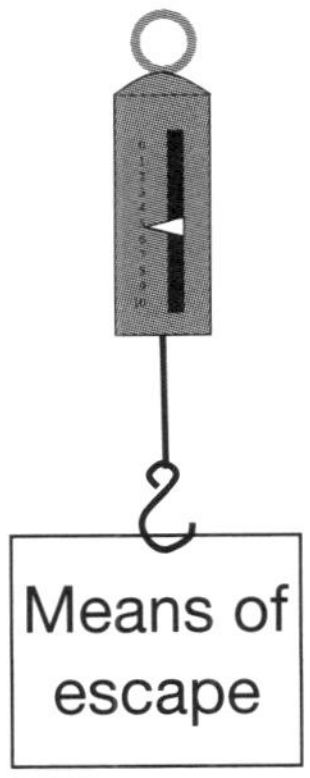

2.4 The design process

2.4.1 General

A framework for a fire engineered approach to building design is described in BS 7974 and illustrated in Figure 2. Clause 4.1 of that code of practice divides the framework into three stages:

- **the qualitative design review (QDR)**, where the scope and objectives of the fire safety design are defined, and where performance criteria are established and acceptance criteria set;
- **quantitative analysis**, where engineering methods are used to evaluate potential solutions; and
- **assessment against criteria**, where the results of the quantitative analysis are compared against the acceptance criteria.

Figure 2: Basic fire safety engineering process (BS 7974)

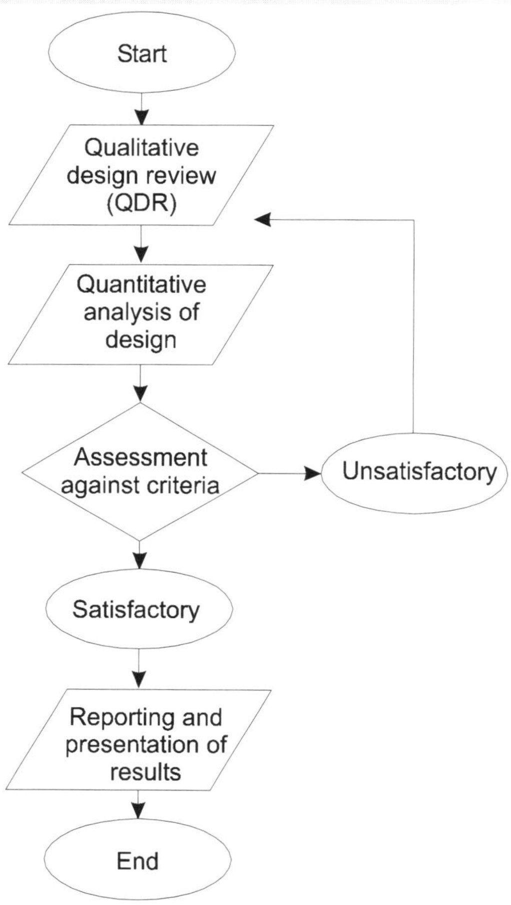

2.4.2 The qualitative design review

2.4.2.1 Personnel involved in conducting the QDR

Attention is drawn to Clause 6.1.2 of PD 7974-0, which states that *'the QDR team on a major project might include'* a representative of the *'approvals body and/or insurer'*, and to Clause 6.2 of BS 7974 which states that *'for smaller projects the QDR may be carried out by a smaller study group'*.

Requirement 1:
To meet the requirements of the insurance industry, wherever practicable, the insurer must be invited to join the QDR team. Buildings are particularly vulnerable to fire during the construction phase, so the invitation must be extended to the insurer of the building during its construction phase or to the principal contractor's insurer.

In some cases, the insurers may not deem it necessary to attend the QDR meeting. In other cases, the owner may not have approached an insurer at the QDR stage of the project. In such circumstances, if the requirements of this document can be demonstrated to have been met, then when selected, the insurers are more likely to be satisfied that the standard of fire safety meets their expectations (although this cannot be guaranteed).

2.4.2.2 Fire safety objectives

Clause 6.3.1 of PD 7974-0 states that *'the fire safety objectives that might typically be addressed in a fire safety engineering study [include] loss control'*. Clause 6.3.3 goes on to describe this particular objective in more detail. It says that *'it might be desirable to take measures to reduce the potential for large financial losses'*. And that *'consideration may be given to minimize the damage to:*
a) the structure and fabric of the building;
b) the building contents;
c) the ongoing business viability;
d) the corporate image.'

Clause 6.3.3 of PD 7974-0 also states that fire safety engineering can be used to assist in an environmental impact assessment of a fire. Clause 6.3.4 goes on to describe this particular objective in more detail: *'consideration may, therefore, need to be given to the limitation of:*
a) the effects of fire on adjacent buildings or facilities;
b) the release of hazardous materials into the environment;
c) methods of fire-fighting (e.g. avoidance of river pollution).'

Requirement 2:
To meet the requirements of the insurance industry, the fire safety objectives of the QDR must include the matters detailed in this document (listed in 2.4.2.2 above) to the extent determined by the agreed acceptance criteria (see 2.4.3).

2.4.3 Acceptance criteria

2.4.3.1 General

Following the quantitative analysis, it is the function of the QDR team to decide whether a satisfactory fire safety solution has been delivered (see Figure 2). In order for this decision to be made without controversy, the QDR team must agree how this decision will be made before the quantitative analysis is started.

Clause 6.4.6.1.1 of BS 7974 describes four methods of judging the adequacy of a proposed solution:

- **Deterministic**, in which the quantitative analysis must show that a predetermined set of circumstances are met. For example, a solution may be considered successful if it can be determined that evacuation will be completed before the ceiling temperature exceeds 100°C.

- **Probabilistic**, in which the probability of a particular event is shown to be acceptably low. For example, a solution may be considered successful if the probability of a fire spreading beyond its room of origin is less than 1:1000.
- **Comparative**, in which the level of fire safety provided by the proposed solution can be shown to be equivalent to that of a similar building which fully complies with approved guidance.
- **Financial**, in which the probability and extent of fire and smoke damage are weighed against the provision of additional fire safety measures in a cost benefit analysis.

PD 7974-0 expands on these criteria in terms of recommendations for loss control objectives. This present document goes further in that it looks at these criteria in terms of *insurer requirements* for loss control objectives.

Figure 3: **Part of an example of a timeline comparison between fire development and evacuation/ damage to property (BS 7974)**

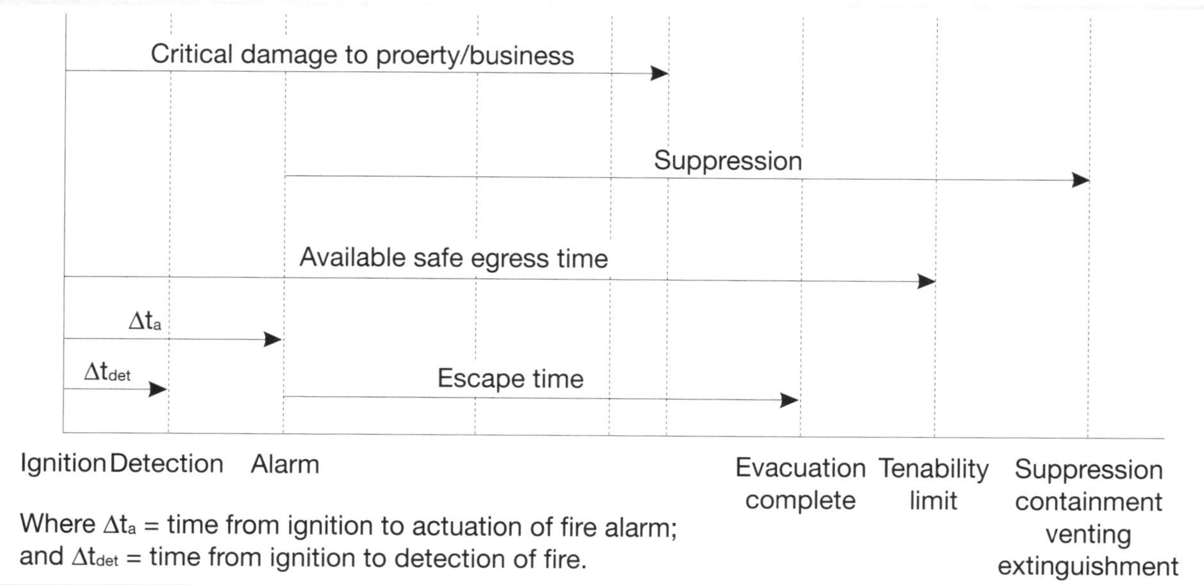

Figure 3 shows part of a timeline comparison between fire development and evacuation/ damage to property. It is taken from BS 7974. This type of diagram could be used to summarise the fact that a set of deterministic criteria have been met – in this case, evacuation is completed before the tenability limit is reached. However, it is reproduced here to draw attention to two things:

- the fact that *'critical damage to property/business'* must be included in any such analysis; and
- that it shows fire causing *'critical damage to property/business'* before suppression, containment or extinguishment occur – indeed, before evacuation has been completed. Although it is accepted that this timeline is only intended as an illustration in BS 7974, it would almost inevitably be unacceptable to insurers as an actual result of a quantitative analysis.

Requirement 3:
In order to minimise insured losses, fire suppression or containment (where used) must be initiated prior to critical damage to property or business taking place.

NOTE: Fire suppression systems must be designed, installed and subsequently maintained in accordance with acceptable codes and agreed with the relevant insurer for the project.

2.4.3.2 Deterministic criteria

It is unreasonable to set property protection criteria for a fire safety engineering design which are any higher than those normally found in a building designed to a prescriptive code. (Ideally, the prescriptive code from which property protection criteria will be taken will be the *FPA Design guide* or the insurer's own standard, but mandatory life safety codes may be used.) In setting deterministic design objectives, it has therefore been assumed that in a building designed to a prescriptive code:

- a 'small fire' (one which does not spread beyond the item first ignited) causes very little damage;
- a 'medium-sized fire' (one which does not spread beyond the room of origin) will damage the contents of the room of origin, but does not cause structural damage; and
- a 'large fire' will cause widespread damage to contents and structural damage.

However, it is feasible for a fire safety engineering design to lack structural fire protection and fire separation. Therefore, what would be a small- to medium-sized fire in a building designed to a prescriptive code, could result in a large fire in a building designed by fire safety engineering. Therefore, stating an acceptable amount of damage for a given fire size is not appropriate.

As an alternative, in this document, determined design targets are based on fire load density. It is assumed that what is normally a small fire remains small because it spreads slowly and is detected and dealt with before it has time to ignite neighbouring materials. A medium fire is larger because the fire load density is larger and fire growth is more rapid. A large fire is catastrophic because the rate of fire spread through a high fire load even outstrips the ability of the fire and rescue service to intervene.

Requirement 4a:
Table 9 of PD 7974-0 gives an example of property protection objectives which may be set as deterministic acceptance criteria. In this document, Table 1 presents objectives which must be met in order to satisfy insurer requirements.

Table 1: **Developing specific performance criteria based on objectives**

Objective	Design target	Performance criteria
Protect the building structure	Where the fire load density is average or below average, ensure that structural damage will not occur as a result of a fire. Where the fire load density is between average and the 80% fractile, ensure that structural collapse will not occur as a result of a fire.	These targets may be achieved in many ways. For example, by limiting the fire load density, by the use of fire suppression, by separation of the fire load to restrict growth or by the use of appropriate structural fire protection.
Limit the loss of the building contents	Where the fire load density is average or below average, ensure that fire damage to contents will be restricted to the item first ignited. Where the fire load density is between average and the 80% fractile, ensure that fire damage to contents will be restricted to the room of origin or to an area not exceeding the maximum compartment size normally allowed for the building in question (whichever is smaller).	These targets may be achieved in many ways. For example, by limiting the fire load density, by the use of fire suppression, by separation of the fire load to restrict growth or by the use of appropriate structural fire protection.
Maintain ongoing business viability	Where the fire load density is average or below average, ensure that a fire will not interrupt normal business for more than 48 hours.(1) Where the fire load density is between average and the 80% fractile, ensure that a fire will not interrupt normal business for more than one week.(1)	These targets may be met by fire protection measures, but they may also be met by appropriate business continuity planning.(2) For example, even if a fire is left to develop to a large size, normal business may be resumed quickly if business critical information is copied and stored off site.(3)

Notes:

If actual figures for fire load density are not available or cannot be calculated, the fire load densities from Table A.19 of PD 7974-1 (Ref. 11) may be used.

The design targets and performance criteria in this table are examples only. It is essential that consultation with the insurer is carried out at the earliest possible stage to ensure that their own specific expectations are addressed.

1. The time periods stated are examples only. Some organisations such as financial institutions and those in the communication sector may incur significant insured losses if their normal activity is interrupted for only an hour or so. In such cases, the 'acceptable' period of business interruption will be much shorter than that stated here.
2. Business interruption can be more costly to the insurer than property loss. Business continuity is therefore a high profile issue with insurers and any fire safety engineering business continuity planning must be agreed with the insurer prior to being adopted into a strategy.
3. Further information on business continuity planning can be obtained from the insurance company that will insure the building on occupation or, if this has not been finalised, from the FPA's *Guidance Note: Business Resilience. A Guide to Protecting your Business and its People* (ref. 12).

2.4.3.3 *Probabilistic criteria*

When setting probabilistic criteria as design targets, it is necessary not only to consider the probability of a fire of a particular size occurring, but also to take into account the damage such a fire could cause.

In a store of inexpensive raw materials, a relatively large fire may have few financial implications. However, the same fire in a store of finished goods could be catastrophic for a business.

PD 7974-0 uses an equation to calculate the Average Potential Loss (APL) of a fire. That equation is reproduced here in a slightly modified form:

$$APL_I = V_I \times DL \times F_{mf}$$

where:

APL_I is average potential insured loss;
V_I is the insured loss resulting from a given fire;
DL is the design life of the building (in years); and
F_{mf} is the frequency of a given fire (in fires per year).

Detailed guidance on the application of probabilistic risk assessment techniques to fire safety engineering is given in PD 7974-7 (Ref. 13).

Requirement 4b:
If a probabilistic approach is used to set acceptance criteria, then the objectives which must be met in order to satisfy insurer requirements must be discussed with the insurer.

In the absence of specific insurer requirements, we recommend the following approach.

For a fire which involves the average fire load density for the building:*

$$APL_I < \frac{\text{annual insurance premium} \times DL}{2}$$

For a fire which involves the 80% fractile fire load density for the building:*

$$APL_I < \frac{\text{annual insurance premium} \times DL}{1.25}$$

* If actual figures are not available or cannot be calculated, the fire load densities from Table A.19 of PD 7974-1 may be used.

An alternative objective which would also be acceptable would be to demonstrate that:

$$(V_I \times DL \times F_{mf}) \leq (\overline{V}_I \times DL \times \overline{F}_{mf})$$

where:

F_{mf} is the predicted frequency of a given fire (in fires per year) in the engineered building;
V_I is the predicted insured loss resulting from a given fire in the engineered building;
$\overline{F}_{mf}$ is the frequency of a given fire (in fires per year) across all buildings of a similar use, type and size; and
$\overline{V}_I$ is the normal insured loss resulting from a given fire in all buildings of a similar use, type and size.

Any probabilistic values or sources of values must be agreed with the insurer during the QDR process. The insurer will wish to be certain that values used are appropriate to the operations and practices of the specific business.

Statistics based on broad generalisations may not be acceptable. Statistics for the frequency of fires in warehouses, for example, may mask a split between a high frequency of fires in warehouses where fork lift trucks are charged and a low frequency of fires in warehouses where they are not.

In applying this objective, it may be determined that insured losses in the fire engineered building would be higher than in a building designed to prescriptive codes, but that this is compensated for by a demonstration that the probability of any given fire size is reduced sufficiently to keep APL_I in the fire engineered building as low as, or lower than, APL_I across all buildings of a similar use, type and size.

2.4.3.4 Comparative criteria

Clause 6.6.2 of PD 7974-0 points out that *'if it is explicitly demonstrated by logical deduction or calculation that the proposed design provides an equivalent or superior level of fire safety to a building complying with an appropriate prescriptive code, the comparative criteria will be met.'*

As such, the use of the comparative approach to setting acceptance criteria represents the most reliable method of demonstrating that a design is satisfactory. This is true whether the design objectives are life safety or building/property protection.

In terms of building and property protection, the insurer requirements for a successful demonstration of comparative success are covered in requirement 4c:

> **Requirement 4c:**
> The degree of structural protection, loss prevention and business continuity assurance provided by a fire safety engineering solution must be at least as high as that provided by a fire safety design for an equivalent building which fully complies with an appropriate code or standard.

2.4.3.5 Financial criteria

Clause 6.8.3 of PD 7974-0 makes the point that when considering the level of fire protection engineered into a building, the cost of installation, maintenance and operational impact may be weighed against the benefit of the added protection. The clause makes this point with respect to probabilistic acceptance criteria, but it is equally applicable to any of the methods discussed in 2.4.3.2 to 2.4.3.4 above.

For example, consider the requirement to ensure that structural damage will not occur as a result of a fire which involves the average fire load density for the building. The implication of clause 6.8.3 of PD 7974-0 is that, if it is calculated that:

total cost of structural damage caused by a fire which involves the average fire load density (direct fire loss, business interruption, loss of corporate image, etc)	= £N
and	
total cost of fire safety measures to prevent such a fire causing structural damage (installation costs, maintenance costs, operational impact costs, etc)	= £2.5N

then as long as no more than two such fires occur during the lifetime of the building, the most cost-effective solution is not to install the fire precautions.

This calculation of whether it is worth investing in fire protection can be carried out in many different ways, with the above being one example. What is important is to view this whole issue from the perspective of the insurer. The insurer is not mentioned in clause 6.8.3 of PD 7974-0, and in an insured building the cost of fire damage and any subsequent business interruption is borne by the insurer. The £N mentioned in the example is the insurance company's money. It is therefore only right that any decision as to the cost-effectiveness of different levels of fire protection should either be made directly by the insurer or by the insured, having fully briefed the insurer and having an accurate indication of the effect on the cost of insurance.

2.5 Quality assurance

2.5.1 Qualification requirements and accreditation

Clause 9.11 of PD 7974-0 states:

'In most fire safety engineering studies, it will be necessary to make some engineering judgements and the expertise of the fire safety engineer will often play a major part in defining the initial design assumptions. To enable a third party to establish that the fire safety engineering study has been carried out by a person with appropriate expertise, the name, qualifications and experience of the individual fire safety engineer(s) responsible for the study should be provided.'

It would be useful to define a particular qualification or accreditation which a fire engineer should hold before being accepted as having 'appropriate expertise'. However, no such qualification or accreditation exists.

Chartered fire engineers who are members of the Engineering Council Division of the IFE have demonstrated competence in fire safety engineering according to its modern definition. However, their numbers are relatively small and they are not bound to follow any particular set of guidelines, other than the competence standards set by the Engineering Council UK, laid down in UK-SPEC (UK Standard for Professional Engineering Competence) and the IFE Code of Professional Conduct.

Equally, graduates of fire science, fire safety management and even fire engineering may have had little or no training in the practical application of the quantitative analysis of fire or the application of fire prevention and protection measures.

There is therefore no obvious way to identify whether or not a particular fire safety engineer has the appropriate skills and experience to undertake a particular job.

2.5.2 Application of appropriate fire safety engineering techniques

As a result of the lack of generally recognised qualifications for fire safety engineers, all that can be required is that:

> **Requirement 5:**
> To enable a third party to establish that the fire safety engineering study has been carried out by a person with appropriate expertise, the name, qualifications and experience of the individual fire safety engineer(s) responsible for the study must be provided. In addition, the fire engineer(s) will follow a recognised set of fire engineering guidelines (such as BS 7974) from the beginning to end of a project.
>
> The fire safety engineer(s) must provide a note of their current professional indemnity insurer, including the policy number and limit of indemnity.

It should be stressed that experience in the application of prescriptive codes is not evidence of competence in fire safety engineering.

2.6 Fire safety management

Good ongoing management of fire safety systems is essential in any building. In a building whose fire safety relies on a fire engineering solution, this is especially so.

In buildings designed to prescriptive codes, a reasonable level of all types of fire safety measures are installed to cope with the majority of expected fires. Thus, in most cases, there is a degree of leeway when it comes to the testing and maintenance of systems, alterations to operations within the building and changes in the fire load.

In buildings designed to fire engineering principles, this may not be the case. Fire safety engineering is often viewed as a bespoke fire safety solution which addresses the specific risks and hazards of the building in question. Alternatively, it could be viewed as a solution which is adequate by the smallest margin. Any alterations to the building or its contents, or any failure in a particular fire safety system, could render the whole solution invalid.

Requirement 6:
To meet the requirements of the insurance industry, the fire engineering solution must take a realistic view of the level of fire safety management which will be applied to the building during its foreseeable future, and must demonstrate that this level of management is sustainable. The fire safety engineer must demonstrate that every effort has been made to gather this information.

Fire safety management requirements such as testing, servicing and maintenance (in particular those aspects upon which the fire safety engineering relies) must be described in the fire safety engineer's report.

If there is evidence that the occupier of the building has a history of good fire safety management, maintaining systems and risk assessing changes to the building environment so that fire safety measures can be updated accordingly, then insurers are more likely to be satisfied with a fire engineered solution which relies quite heavily on robust fire safety management. If no such evidence exists, or if there is evidence to the contrary, then insurers will expect a fire engineering solution which is robust in itself.

2.7 Issues associated with the construction phase

Buildings are at their most vulnerable to fire during their construction phase. There is often a high fire load present as building materials are stored on site prior to installation. Hot work may be carried out, there may be a multitude of potential ignition sources as contractors use electrical equipment in varying states of repair, and poor site security can lead to arson attack.

In a building whose fire safety is designed to a prescriptive code, a fire which breaks out during construction can have a devastating effect prior to the completion of fire separation and fire protection of the structure. In a building whose fire safety is designed using a fire engineered approach, this situation may be even worse. If the intention is for fire safety to rely on automatic detection and sprinklers, rather than fire separation and protection, then the building is at great risk from fire, right up until the commissioning of the fire safety systems.

Requirement 7:
Fire safety management of construction sites is always important, but to meet the requirements of the insurance industry, where the fire safety design of a building is engineered, then a risk assessment of the implications of the fire engineering to the construction phase fire risk must be carried out and must be available to the insurer.

2.8 Reporting and presentation

According to clause 9.1 of PD 7974-0, the format of a fire engineer's report will depend on the nature and scope of the study, and on the style of the individual engineer. However, the clause specifies that 'typically' the report should contain the following information:

- objectives of the study;
- building description;
- results of the QDR;
- quantified analysis;
- comparison with acceptance criteria;
- fire safety strategy;
- management requirements;
- conclusions;
- references; and
- qualifications and experience of the fire safety engineer.

Requirement 8:
With respect to the issues addressed by this document, the fire engineer's report must contain a non-technical explanation of how requirements 1 to 7 have been met. In the case of large or complex projects, it may be necessary to produce this information as a separate report for the insurer.

The information required to be contained within the fire engineer's general report, or within the report produced specifically for the insurer, is described in the following section: 'Information required by the insurer'.

Information required by the insurer

Integration of insurer requirements (Requirement 1)

Explain how the requirements of this document were integrated into the design process, including:

- confirmation that insurer interest was raised with the client; and
- description of the level of insurer involvement at QDR stage, either:
  - application of this document with direct insurer involvement; or
  - application of this document following contact with insurer; or
  - application of this document without contacting insurer.

Property protection fire safety objectives (Requirement 2)

Confirm that the property protection fire safety objectives of the project include at least:

- the structure and fabric of the building;
- the building contents; and
- the ongoing business viability.

Business criticality statement (Requirement 3)

Confirm that preventing business-critical property damage has been given a sufficiently high priority.

Property protection acceptance criteria (Requirement 4)

Describe the type of acceptance criteria that have been selected for property protection issues (deterministic, probabilistic or comparative), and explain the reason for the selection.

Comparison against acceptance criteria (Requirement 4)

Describe specific objectives within the particular acceptance criteria selected, and explain how those objectives have been met.

It is important to stress that non-technical descriptions are required in the report, although these descriptions must be backed up by quantitative analysis which can be assessed by a third party if it is deemed necessary.

Describe specific design targets and discuss performance criteria within the particular design objectives.

Qualifications and experience of the fire safety engineer(s) (Requirement 5)

Provide the name, qualifications and experience of the fire safety engineer(s) responsible for the study.

Fire engineering guidelines followed (Requirement 5)

Cite the recognised set of fire engineering guidelines that have been followed. If the guidelines have been deviated from in any way, explain the reason for the deviation.

Interaction of fire engineering with fire safety management (Requirement 6)

Describe the predicted level of fire safety management that will be applied to the building on occupation.

Explain the basis on which the prediction above is founded.

Describe how the chosen fire engineered solution has taken account of the predicted level of fire safety management with respect to property protection issues.

Impact of fire engineering on construction phase (Requirement 7)

Describe the impact of the fire engineered solution on fire safety during the construction phase of the project, and explain how adverse effects are to be managed.

References

1. BS 7974: 2001: *Application of fire safety engineering principles to the design of buildings. Code of practice,* British Standards Institution.
2. *FPA Design guide for the fire protection of buildings: Essential principles,* 2nd edition, Fire Protection Association for InFiReS, 2007.
3. *CIBSE Guide E: Fire engineering,* the Chartered Institution of Building Services Engineers, 2003.
4. PD 7974-0: 2002: *Application of fire safety engineering principles to the design of buildings. Guide to design framework and fire safety engineering procedures,* British Standards Institution.
5. The Building Regulations 2000 (as amended), Statutory Instrument 2000 No. 2531, The Stationery Office, 2000.
6. The Building (Scotland) Regulations 2004 (as amended), Scottish Statutory Instrument 2004 No. 406, The Stationery Office, 2004.
7. The Building Regulations (Northern Ireland) 2000 (as amended), Statutory Rule 2000 No. 389, The Stationery Office, 2000.
8. Approved Document B to the Building Regulations 2000: *Fire Safety (Volume 2): Buildings other than Dwelling Houses* (2006 edition), NBS for the Department for Communities and Local Government, 2007.
9. *Scottish Building Standards Technical Handbook (Non-domestic),* Section 2: Fire, The Stationery Office, 2008.
10. Technical Booklet E to the Building Regulations (Northern Ireland) 2000: *Fire Safety,* Department of Finance and Personnel for Northern Ireland, 2006.
11. PD 7974-1: 2003: *Application of fire safety engineering principles to the design of buildings. Initiation and development of fire within the enclosure of origin (Sub-system 1),* British Standards Institution.
12. *Guidance Note: Business Resilience. A Guide to Protecting your Business and its People,* Fire Protection Association for InFiReS, 2005.
13. PD 7974-7: 2003: *Application of fire safety engineering principles to the design of buildings. Probabilistic risk assessment,* British Standards Institution.

B
Index

C

E

G

H

Q

R

S

T

U

V

W